Electrical Handbook For RVs, Campers, Vans, Boats, & Trailers

Electrical Handbook For RVs, Campers, Vans, Boats, & Trailers

by Herb Gill

TAB BOOKS
BLUE RIDGE SUMMIT, PA. 17214

FIRST EDITION

FIRST PRINTING—APRIL 1979

Copyright © 1979 by TAB BOOKS

Printed in the United States of America

Reproduction or publication of the content in any manner, without express permission of the publisher, is prohibited. No liability is assumed with respect to the use of the information herein.

Library of Congress Cataloging in Publication Data

Gill, Herb.
   Electrical handbook for RVs, campers, vans, boats & trailers.

   Bibliography: p.
   Includes index.
   1. Recreational vehicles—Electric equipment. 2. Vans—Electric equipment. 3. Boats and boating—Electric equipment. 4. Storage batteries. I. Title.
TL272.G53   629.2′54   78-12692
ISBN 0-8306-8867-5
ISBN 0-8306-7867-0 pbk.

Cover photo courtesy of LRP.

Preface

This book was written because of the questions that have been asked by fellow recreation vehicle owners about how to install electrical equipment in them. As the owner of an outboard cruiser, a self-contained travel trailer and a four-wheel drive vehicle, I have had some experience in installing and using special electrical circuits and equipment. As an electrical engineer, with considerable test equipment available, some of the installations have been duds no less. So it is hardly to be expected that a person, inexperienced in electricity, will have much luck in these special installations, particularly since there are few guidelines and little published data to go by.

Several of the duds were the result of making changes without analyzing the possible results. Some of the duds were the result of insufficient knowledge of what was needed or necessary. This book is an attempt to bring out some of these hidden problems and how to avoid them. My friends' questions demonstrate that others need to know the answers also.

One of the most difficult decisions to make in preparing the book is how far to go into the basic principles of electricity, and when to stop before getting too technical. With over 30 years of experience in electrical work, it is hard to avoid going into the subject over your head since the material is practically second nature to me. At the same time, there is little point in making the information so bland that it is of no use to anyone. I hope I have been able to steer a middle course.

There are quite a number of books on the market today that teach the elements of electricity—that will not be covered—but a chapter is devoted to definitions and explanations of some of the basics necessary to cover the circuits in the book.

One big problem with understanding electricity is that it is invisible, yet it can make itself shockingly evident like when you grab a spark plug terminal. Also, too much in the wrong place causes sparks, smoke, and fire.

If you want to learn about electricity, work with it. Stay away from the spark plugs or coil; the 6 or 12 volts cannot shock you. If you are afraid of the smoke and fire, just connect a fuse in the circuit you are trying; if it is wrong, only a fuse is lost. Study the circuit, put in another fuse and try again. That is the way to learn.

Do not be afraid of the simple mathematical formulas given in the book.

The only way to know what electricity is doing in a circuit is to measure it. Even an expert electrical engineer must have instruments; he cannot see the stuff either. The few instruments needed for recreational vehicle (RV) electrical testing cost about as much as a half-gallon of whiskey; after you have completed your first successful circuit though, the half-gallon jug may be necessary to celebrate!

This book attempts to be a practical working handbook; there is little theory and little explanation of how electricity works. There are some explanations of why a particular circuit or device is used, and some advice on what not to do. Practical data is given in as much detail as is available to the author at this writing. Some of it is woefully inadequate. It is to be hoped that manufacturers of popular items will furnish more specific data for their products some day rather than inflated advertising claims. But this might invite competition, since there is considerable room for improvement of efficiency and performance in many of these products.

When the amount of energy is limited, as is true of a battery source, it is important that a device be efficient. My first experience with this problem was a dead starting battery in a boat 1 mile off the beach. The bait pump had run down the battery, and there was no way to crank the engine by hand. I have to thank the U.S. Navy for an able assist. Their umpteen-horsepower torpedo boat towed my little one so fast, the propeller easily turned over and started a cold, dead engine.

On two other occasions, I had to help out fellow boaters with the same problem. Out came the electrical instruments—the bait pump current was measured at 150% of the rated value. This drain would kill any healthy auto battery in 6 or 7 hours. Mine were not healthy anyhow, so I heeded the meters answers and changed the pump and battery.

Another dud was an attempt to charge a battery in a trailer from the generator of the towing vehicle. The hookup didn't work. A few meter readings told why.

A friend put a diode isolator in his camper for a second battery in the cabin. In about 1 week, he couldn't start his truck. He bought a new battery; 2 weeks later it was the same story. Meters again showed why.

As a result of these experiences and others, the engineering of some of the special circuits used in campers, boats, and other recreational vehicles has been worked out. The circuits are shown, and the checks necessary to test them are outlined. Incidental to this work, considerable data was assembled that has not been given in a popular book. It is available in engineering and science literature if one knows how to dig it out; what is applicable to these problems is given here.

The circuits have been checked for proper operation, but they will be in trouble if an incorrect connection is made. Also, some may require a balancing of loads and circuit resistances to be fully satisfactory.

All the answers cannot be printed in a book; there are too many possible problems. But, a starting point is presented from which you will have to run your own race.

Again, I must say that I hope I have not been too technical so the information cannot be used. But, electricity is not a simple thing, so some study of it is required for you to make it do your bidding. Above all, do not be afraid to experiment!

I particularly wish to thank my friend, Jim Barry, for his encouragement and suggestions for subjects to be covered. Also, I must thank my wife, Virginia, for her forbearance in not getting some of her problems solved while this book took up all of my time.

Herb Gill

Contents

1 What It's All About 11

2 Charging Circuits 15
Class 1A—Class 1B—Class 2—Class 3—Class 4

3 Second Battery Hookup Methods 19
Method A-Direct Paralleling—Method B-Manual Switch—Method C-Isolating Relay or Contactor—Method D-Diode Isolators—Method E-Twin-Engine Systems

4 Problems of the Second Battery 31
Twin-Engine Systems—Dual Battery Recommendations

5 Electrical Terms You Should Know 41
Measuring Electricity—Examples—Electrical Measurements—Electrical Wiring Diagrams and Schematics—Relay and Switch Contacts

6 Battery Basics 57
Construction—Measuring Battery Condition—Battery Charging—Maintenance & Repair—Battery Water—Battery Capacity—Automotive 24 Super—Automotive 24 Standard & 22 Super—Automotive 22 Standard—6-Volt Automotive 1 Super—Estimating Battery Loads—System Checks—Automotive Batteries Point-Voltage Readings

7 Generators & Alternators 75
Principles of Operation—High-Voltage Output

8 **Wire Sizes** 81
Battery Charging Wires

9 **Fuses & Circuit Breakers** 89

10 **Electrical Loads** 93
Edison Screw Base (Household Lamps—Fluorescent Lights With DC-AC Converter—Fans—Pumps—Motors—Radios, Winches—DC-AC Inverters—Electric Trailer Brakes—Inrush Current—115-Volt Alternating Current Loads—Automotive & Marine Light Bulbs—12-Volt Type

11 **Special Boating Problems** 99
Running Lights—Wire Terminals

12 **Electric Trailer Brakes** 105

13 **High-Current Charging** 113

14 **Auxiliary Power Units** 123

Appendices 132

Bibliography 139

Index 140

Chapter 1
What It's All About

One of the most common questions asked about the electrical system of recreational vehicles is how to install a second battery in such a way that the camper electrical loads will not run down the vehicle starting battery. A dead starting battery can be a matter of life and death for a boat at sea or out of touch with normal traffic on a river or lake. Also, a camper in an isolated spot in the summer desert must have a good battery to start his engine or use an emergency radio transmitter.

Less critical but important questions are: "How long will my battery last if I use five 21 candlepower lamps all evening in my cabin?", "What size wire should I use to connect all the lamps?", "What size fuse is right for this circuit?"

These question can be answered by anyone with only a limited knowledge of electricity if he is shown how to do it. For those who need this little bit of electricity, a chapter in this book is included to help those willing to learn. The subject is not too difficult if you are not scared of electricity. Just because it cannot be seen but can be felt is no reason to be that way about it!

In order to solve electrical problems properly, it is very desirable to have several test instruments. These need not be very expensive. One convenient one is a small volt-ohm-milliammeter, VOM for short. Get a low-resistance type

(1000 ohms per volt) with at least one range of 15 to 20 volts. (In the Appendix is a list of possible sources for all materials and parts listed in the book.) Also, a zero-center ammeter with a full-scale deflection of 30 amperes is good. There are now available ammeters which need only be held against the wire to read the current flowing. Most have two ranges, 0 – 60 and 0 – 400 amperes. The latter scale is used to check starter motor currents. The hold-on types are not as accurate as the in-circuit types, but are quite adequate. Several of these meters are shown in Fig. 1-1.

Another simple but effective tool is a 12-volt light bulb soldered to a pair of 3-foot wires. This homemade tool is often quicker and easier to use than a meter for checking the presence of voltage on a circuit.

To return to the question on batteries. There is no simple way to connect two batteries in any vehicle so the starter battery is protected from the other loads. The problem is that there are several different charging systems used in cars, trucks, and boats. Also, there are several different methods of hooking up the two batteries.

One method of adding a second battery is simply to connect it to the load in the vehicle, and take it out at the end of the trip and recharge it at home. This method is not discussed in the book, but the chapter on storage batteries should be read since keeping such a separate battery well maintained can be a problem.

One chapter is devoted to the description and classifying of the various charging systems that are normally found in American-made vehicles. Another chapter describes the various methods of connecting the batteries. A third combines the systems and describes possible problems and their solutions.

For the newcomer, there is a chapter on basic electricity and the reading of electrical diagrams. It is not in too much detail, but there are plenty of basic books on this subject that can be referred to.

Generators and regulators are covered to the extent they affect the RV battery problem. Battery ratings, calculating loads, time of operation, and the influence of outside factors on battery performance are covered in some detail.

Fig. 1-1. Various meters. Top left—ammeter shunt used to increase the range of high-current ammeters. Top center—volt-ohm-milliammeter (VOM) used for testing circuits. Top right—panel type DC voltmeter with a range of 11 – 16 volts. Center left—aircraft-type ammeter with a 0 – 80-ampere range. Center right—panel-type DC voltmeter with a 0 – 15-volt range. Bottom left—auto-type ammeter that reads –30 to +30 amperes. Bottom center—special ammeter that indicates starter and generator current. Bottom right—panel-type DC ammeter that indicates –20 to +20 amperes.

Wire sizes, current capacity, voltage drop calculations and special problems related to RVs, including fuse sizing fill one chapter.

Some special problems like bait pumps for boats, trailer brake adjustments and high charging rate circuit end the book.

An appendix lists a number of possible sources of parts and materials described in the book. Trailer and RV stores are not listed as most are local. Local radio and electronics parts houses, auto supply stores and marine hardware stores are possible sources also. One may have to go to electrical wholesalers for high current switches, fuses and circuit breakers. Do not overlook the large mail order companies like Wards or Sears.

The data given is restricted to that most used for autos and RV applications. For the most part, 12-volt units are described unless specified otherwise.

Current drain data are approximate—measure the unit you have and be sure, or ask the manufacturer to supply exact data.

Chapter 2
Charging Circuits

In this chapter common charging circuits used in automobiles, small trucks, and boats are classified into types. This classification is necessary because some dual-battery hookups are not satisfactory with all of these charging circuits.

All charging circuits consist of an engine-driven generator or alternator and a voltage regulator—in some cases they may include reverse current and current limiting regulators. Some circuits also use a field relay that cuts off the field current of the generator when the engine is stopped or the ignition switch is turned off.

A fundamental difference in the systems, as far as dual-battery hookups is concerned, is where the battery voltage is sensed. If the sensing point is outside of the voltage regulator, there are more possibilities of dual-battery connections. A sensing point within the regulator limits battery hookups to one type, unless you are willing or able to change the regulated voltage setting or to rebuild the regulator. Let's look at how these systems are classified.

CLASS 1A

The oldest system used on vehicles is that of direct current (DC) generator with a regulator having three units. These are the voltage regulator, the current regulator, and the reverse-current relay. All these systems use a voltage sensing unit within the regulator. The system voltage is rated at 6 volts (3-cell battery), with a minor variation using 8 volts (4-cell battery). This class uses either a positive or negative ground system, with most being negative ground.

CLASS 1B

Historically, the 12-volt DC system was used next. The circuit and components are identical to the Class 1A system except for the voltage change. Practically all of these are negative ground systems.

The regulator for both of these classes is a rectangular metal box with three terminals on the base, usually marked A or ARM, B or BAT, and F or $FIELD$. Total generator current flows through the regulator.

Most of these regulators can be adjusted for voltage, but in some the job is not easy. Popular auto manuals cover the procedure.

CLASS 2

Class 2 systems are currently being used on all American-made cars and most foreign models. The generator is actually an alternator with electronic diodes being used to rectify the alternating current (AC) output to DC for charging the battery. All systems have a negative ground except a few early outboard motor systems.

There is considerable variation in the hookup of the alternator and regulator. There are numerous variations of the regulator internal construction, including mechanical vibrating relays and wholly or partially transistorized units. Some variations in this class place some components in different boxes.

The basis for the classification of subsystems in this class is where the voltage-sensing point is connected.

Class 2A

The simplest AC system and the easiest to identify is the single-unit, mechanical-vibrating-relay type used by Chrysler and Prestolite. The regulator box has only two terminals, one marked F, connected to the alternator *field*, and one marked I, connected to the *ignition* switch. The voltage is sensed where the I terminal is connected. Current flows through the regulator and field whenever the ignition switch is on.

Class 2B

This is the class used by Ford and most Delco (General Motors) units, but not all. There are two units in the regulator box, a field relay and the voltage regulator. The field relay is closed when the ignition switch is turned on and connects the alternator field to the voltage regulator contacts. Voltage is sensed at or very close to the battery terminals by a separate lead.

The Ford regulator terminals are F to the field, S to the generator stator centerpoint, $+A$ to the battery, and I through the charge indicator light to the ignition switch.

If an ammeter is used rather than a light, terminal S goes to the ignition switch and terminal I is not used. Terminal S of the generator is also not used.

Delco has two 4-terminal regulators that look alike; one is a 2-relay type and the other is partially transistorized. The terminals are marked BAT (to the battery), $F1$ and $F2$ (to the alternator field), and SW (to the ignition switch) in the earlier models. In the later models, the terminals are marked F (to the alternator field), 2 (to the alternator field marked R), 3 (to the battery terminal), and 4 (to the ignition switch through the charge indicator light).

In the earlier model, the ignition switch activated the field relay directly in order to connect the voltage regulator to the battery. In the later models, the voltage to close the relay comes from the alternator. This means it has to be running before the relay will close, and if the alternator stops for any reason, the relay opens. The voltage setting of these regulators can be changed.

Delco has a fully transistorized unit in this class that has three terminals marked POS, NEG, and FLD. It has a separate field relay in another box. Terminals of this relay are marked 1, 2, and 3. Number 2 is the voltage-sensing lead, usually going directly to the battery. The relay has a sheet-metal box, but the regulator is in a cast-aluminum housing. The voltage of this unit is adjustable.

Motorola systems are class 2B units but the voltage-sensing lead is connected to a special isolating diode rather than directly to the battery. The voltage regulator is set to compensate for the voltage drop in the diode, so the effect is the same as measuring the voltage at the battery. The voltage of this unit cannot be changed.

CLASS 3

Class 3 systems are alternator powered, but the regulators are similar to the older DC generator units in that the voltage is sensed within the regulator.

The Leece Neville unit is of this type, with a 4-terminal regulator having terminals labeled *BAT, GEN, FLD*, and (on the opposite side of the case) *IGN*.

The Prestolite system with a charge-indicator light has a 5-terminal regulator labeled *A* (alternator output), *I* (ignition switch), *L* (indicating light), *B* (battery), and *F* (alternator field).

The newest Delco self-contained unit with an integrated circuit (IC) regulator built into the alternator is Class 3.

There are some uncommon early Delco units having a 4-terminal regulator with three relays, a reverse-current cutout, a field control, and a voltage regulator. They look like the early transistor regulators but the center of the three terminals goes to the generator output rather than to the F1 terminal as do the transistor units. The other terminals are alike on all units. Watch for this one!

CLASS 4

This class covers small alternators with a rectifier but no voltage regulator. Most outboard engines are made this way. The class also covers most high-frequency AC generated and diode-rectified DC auxiliary power plants.

The highest charging current is controlled by the top speed of the engine and the resistance of the alternator windings. The charging current varies with engine speed and battery condition.

Chapter 3
Second Battery Hookup Methods

There are several ways of hooking up the second or auxiliary battery; each has its advantages and disadvantages. Some circuits cannot be used with certain classes of charging circuits described in *Chapter 2*.

METHOD A—DIRECT PARALLELING

Simply hooking two batteries in parallel will increase the capacity of the system. The method is simple and will work with all classes of charging circuits. There is, however, no way to prevent both batteries from being run down by a load. The interconnecting wires must be heavy enough to share the starter current or they will burn up. Both batteries should be of similar construction and age, but they do not have to be the same size. By construction is meant plastic or rubber case, top strap or internal cell connections, and regular or high voltage. The importance of age is that an old, leaky battery will run down its companion too if they are connected in parallel.

METHOD B—MANUAL SWITCH

If a manual switch is connected between the two batteries (Fig. 3-1), the secondary one can be isolated from the starting battery at any time. When the switch is closed, both of the batteries will be charged. If no attempt is made to use the secondary battery for starting, the switch need only carry the charging current. However, if starting is contemplated with both batteries, or where there is danger of forgetting to open the switch while starting, the device must be able to carry the shared starting current. This would mean a 200-ampere rated switch with AWG #4 or #6 wire. Mounting such a unit within convenient reach of the driver can be awkward.

About the only place it is possible to buy manually operated switches of this rating is in a marine hardware store. They are used to open the battery circuit when the boat is moored.

Any switch used to interconnect batteries must be capable of handling at least 50 amperes, since when a well-charged battery is paralleled with a discharged one, the full one will discharge into the other until both are equalized. Never attempt to parallel a dead battery with a full one unless a 200-ampere switch is used.

METHOD C—ISOLATING RELAY OR CONTACTOR

The switch mentioned in *Method B* (Fig. 3-1) can be suitable relay or contactor that may be mounted in any convenient place. Only small-gauge control wires need be run to the driver's station for control. A small switch can be used to control the contactor, or it can be connected to a terminal of the ignition switch. In this manner the relay is closed only when the ignition switch is on. Do not connect the control wire to the accessory terminal because the relay will be turned on with the accessories, which will defeat the purpose of the relay system.

There are a number of relays available for this application. Some engine starter relays can be used if their coil is rated for continuous-power application. If one is available, but the duty is unknown, it can be checked by applying 12 volts to the coil. If the current draw is more than 0.75 ampere, the relay should not be used since it may overheat and burn out. The coil resistance also can be measured. If it is less than 9 ohms, it is taboo.

One relay specifically designed for this type of service with continuous coil power is the RBM Model 70, made in either 6- or 12-volt versions.

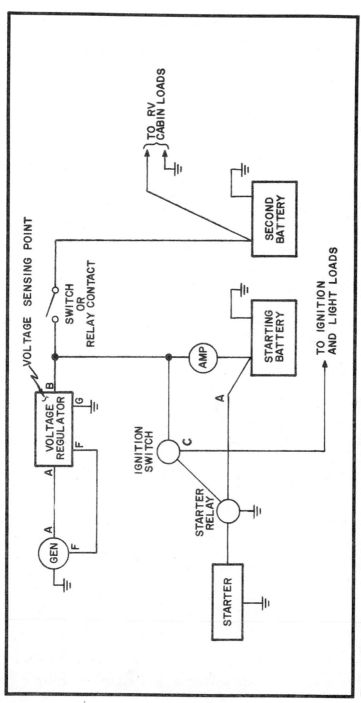

Fig. 3-1. Wiring diagram for Methods B and C. It is used with Class 1, 2, and 3 charging circuits. The switch for the second battery can be operated manually or by relay.

It is sometimes possible to use a surplus 24-volt aircraft relay, particularly if it is rated for intermittent 24-volt duty. Some of these relays have two coils, a low-resistance one to pull the armature in, and a high-resistance one to hold it. These are excellent for paralleling service, but the resistance check of the open relay should not be used to decide its worth. The closed-contact, 12-volt current drain is what counts.

In Fig. 3-2 are shown several typical relays and contactors. Upper-left example is a double-pole, double-throw 15-ampere contact rating unit, typical of heavy-electronic or light-power relays. Normally, these relays are not enclosed, as they are used in a larger control cabinet. Such a relay should never be used under the hood of a car without being enclosed; dirt and oil will get into the contacts and cause faulty operation.

Lower-left one is a 3-pole, double-throw 5-ampere contact relay typical of what is used in electronic control circuits. These are also available in a plastic case.

Upper-right example in Fig. 3-2 is a 200-ampere 2-coil aircraft relay. Note the plastic case on the bottom containing the coil-transfer switch. Although the coil is rated for 24 volts, these 2-coil relays operate well on 12 volts.

In the center is a typical heavy-duty automotive starter contactor rated at 200 − 300 amperes. Of the two smaller screws, the one on the left is the ignition-resistor-bypass terminal, and the one on the right is the coil terminal. The other terminal for the coil is tied to the mounting bracket. The bypass terminal is not used. The RBM relay mentioned earlier is similar to this, but the coil is attached to the small terminals and there is no resistor-bypass terminal.

Do not connect the relay to the ignition-coil terminals; there may be a resistor in the lead between the coil and the ignition switch. Having the relay on the coil side of this resistor will decrease the voltage enough to cause ignition troubles. Connect the relay to the ignition switch only.

Use heavy wire for the relay to battery connnections— not less than AWG #10 for normal use or regular starter cable if you intend to use the second battery for starting. Of course, using a battery located 10 feet away in a trailer for starting is impractical. Large wire here may be required for another

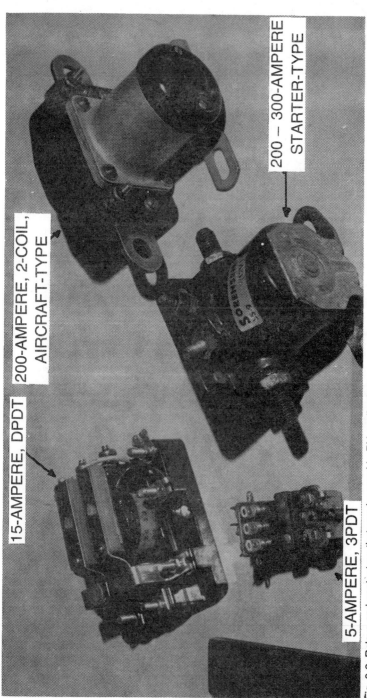

Fig. 3-2. Relays and contactors that can be used in RV applications.

which will be discussed later in the chapter on deter- correct wire sizes.

D D—DIODE ISOLATORS

The latest method of connnecting two batteries is employing a diode as an isolator. The diode is, in effect, an electrical check valve. It will allow electricity to flow through it in one direction but will block it in the other. Hence, a diode placed between two batteries will allow current to flow from one to the other but will block the flow from the second back to the first. The diode is not a perfect valve, however, since there is always some voltage drop or loss when the current is flowing in the forward or conducting direction. This amounts to between 0.2 and 1.0 volts, depending on the design of the device and the amount of current flowing.

The effect is not the same as a resistance in the line, however, because there is a minimum voltage which must be exceeded before any current at all can flow. In normal resistance, there is no limit to how small a voltage is needed to cause some current to flow. Also, the reverse-blocking effect is not 100% perfect, although in modern units the amount of reverse leakage is almost negligible.

The forward voltage drop of a diode is a very important property in isolating two batteries which must be charged from the same power source. The batteries are very sensitive to the charging voltage and only a small change—in the order of a few tenths of a volt—will make the difference between an undercharged and an overcharged battery. When two batteries are charged in parallel, they are actually being charged for exactly the same time, so a difference in voltage between them will be noticeable in the amount of charge each receives. Also, batteries are sensitive to temperature, requiring a higher charging voltage as the temperature goes down.

The voltage regulator in the car is purposely designed to charge a battery harder if the air temperature around the regulator is low. So, if the battery is not at about the same temperature as the regulator, the charging rate can be upset.

The diode voltage drop may interfere with the proper charging of a battery if the diode is inserted between the battery and the regulator voltage-sensing point; hence, the

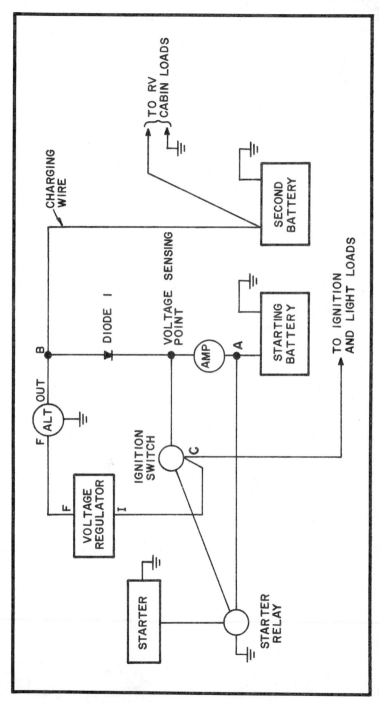

Fig. 3-3. Wiring diagram for Method D1 in a Class 2A system.

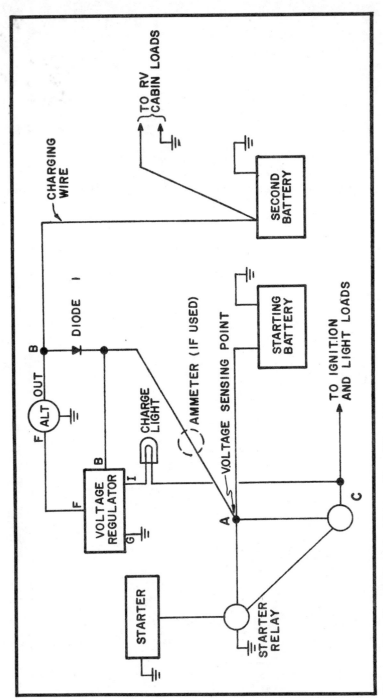

Fig. 3-4. Wiring diagram for Method D1 in a Class 2B system.

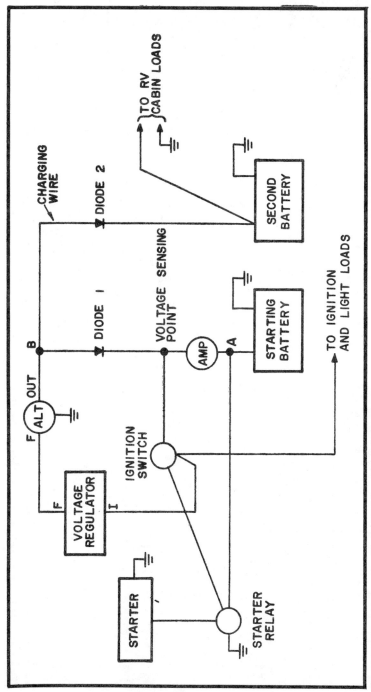

Fig. 3-5. Wiring diagram for Method D2 in a Class 2A system.

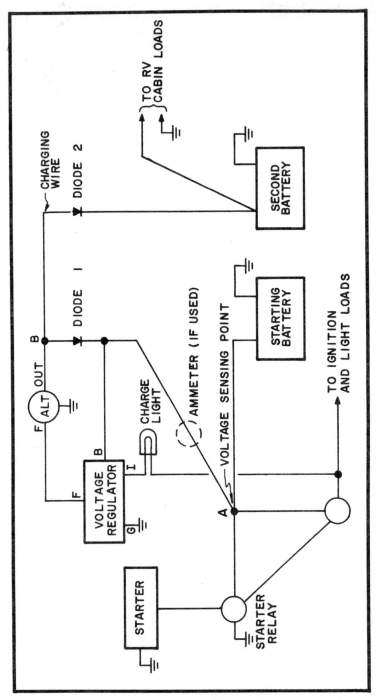

Fig. 3-6. Wiring diagram for Method D2 in a Class 2B system.

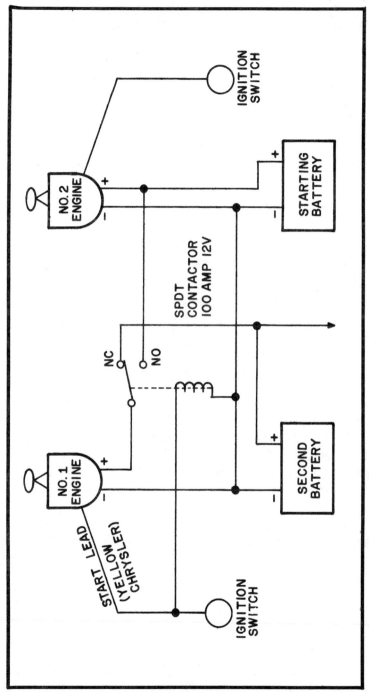

Fig. 3-7. Wiring diagram for Method E in a Class 4 system.

emphasis on locating this point in the description of the charging systems.

On the other hand, the diode voltage drop may not affect the system enough to be a problem, and it can even be used to advantage to improve the overall performance.

Two versions of the diode system are useable: Method D1 uses only one diode in series with the starting battery (Figs. 3-3 and 3-4). The second uses a diode in both battery charging leads. This is method D2. One diode only in the auxiliary-battery lead cannot be used (Figs. 3-5 and 3-6).

METHOD E—TWIN-ENGINE SYSTEMS

In marine installations, where there are often two engines installed, another system can be used. This system is almost a necessity for small outboards where the available charging current is low.

In this system, one engine is connected to one battery that is used for starting only, and the other engine is connected to the secondary battery that serves the lighting and cabin loads. When this engine is to be started, a single-pole, double-throw switch or contactor is shifted to connect this engine to the starting battery. After the engine is running, its charging output is shifted back to the secondary battery. A wiring diagram for this circuit is shown in Fig. 3-7.

A suitable contactor for this operation is the RBM Type 70-910, with SPDT (single-pole, double-throw), 100-ampere contacts, and a 12-volt continuous-duty coil. This relay is similar to the one in the center of Fig. 3-2, but it has two large contact studs on the top.

It is also possible to connect this circuit using two single-pole, normally open contactors such as the RMB 70-902.

Chapter 4
Problems of the Second Battery

As has been hinted, there is more to hooking up a second battery than appears at first look. When the two batteries are close together, are nominally the same temperature, size, and age, with a very low-resistance wire between them, there is no problem. A switch or relay can be used to isolate the two during the discharge cycle, and to connect them together for charging. The switch satisfies the low-resistance needs of the paralleling circuit very well, as does a contactor or relay.

Figure 3-1 shows the wiring to use with any type of vehicle charging system if a switch or contactor is used to parallel the two batteries. The diagram is shown for the Class 1A, 1B and 3 systems, for which it is ideal, but it can be used for any other type. If a contactor is used, the coil should be connected at point C on the ignition switch. If the second battery is on a trailer, the connecting wire should be as heavy as practical.

One diode alone in the starter battery charging wire will satisfy the isolation requirements of the two batteries. This circuit for a Class 2A system is shown in Fig. 3-3, and for the Class 2B system in Fig. 3-4. There is not a great deal of difference between these two circuits; the principle problem is to locate the points to connect the wires, since there are variations between vehicles as to how the wiring is routed.

Reference to the wiring diagram for the particular vehicle will help.

How well this system works as far as charging the batteries depends on the circuit resistances between the batteries. There is one combination of circumstances where the system will work quite well. With the voltage regulator connected to point A, sensing the actual starting battery voltage, it will require the alternator to put out a voltage necessary to charge the battery plus the voltage drop in the diode in the charging line. Now the auxiliary battery is connected to the point B, directly at the alternator output. If the voltage drop through a relatively long, small wire to this battery is about equal to the diode voltage drop, both batteries will charge about equally. If, however, the interconnecting wire is a short, heavy one, the RV cabin battery will be overcharged because it will have an excess voltage on it when the starter battery is fully charged.

For example, if the proper voltage to charge the starter battery is 13.9 volts, and the diode drop is 1 volt, the generator will be called on to put out the sum of these or 14.9 volts in point B. Now, if the second battery is in a trailer where the colder temperature requires a charging voltage of say 14 volts, then with two 35-foot sections of AWG #14 wire running to and from the second battery, the voltage drop at 10 amperes is about 0.9 volts. This will provide the proper charging voltage.

But, in the case of a van camper, where the second battery is right alongside of the starter battery, the secondary battery would be overcharged because it would have almost 14.9 volts applied to it when it only needed 13.9 volts, the same as the starter battery. In an extreme case, under these conditions the secondary battery could absorb all of the current from the alternator, and the starter battery would not charge at all.

If a second diode is added in series with the secondary battery-charging lead, as shown in Figs. 3-5 and 3-6, some of these problems are answered. Since there is a diode in both battery leads, the charging will be balanced for the systems where the batteries are close together. A battery at the end of a long wire in a trailer will be at a disadvantage, however, and will tend to be undercharged.

This dual diode isolator is the kind being sold by several companies today. The diodes are connected together at the alternator, usually by an internal connection in commercial units.

In making the connections of Figs. 3-5 and 3-6, be careful to have the voltage regulator sensing lead on the starting-battery side of the diode. If the voltage is sensed inside of the regulator as is done in Class 1A, 1B and 3 systems, the diode must be placed between the battery and the sensing point so that there will still be problems. The voltage on both batteries will be too low if this is not done by the amount of the diode voltage drop. And, if the one is in the trailer, the resistance of the long charging wire will further decrease the input to that battery.

With a switch or contactor, the two batteries can be equalized over a period of time if the switch is kept closed, even though the engine is not running. But with diodes, this equalization cannot occur.

There is one possible cure to the diode drop problem with Class 1 and 3 systems. The setting of the regulator can be increased to compensate for the diode drop. The method of doing this is described in a number of automotive handbooks.

When the secondary battery is in a trailer, where a long charging wire must be run to it, for all systems except the single-diode method, the charging wire should be as large as possible to minimize voltage drop. Also, particular care should be used to have a good ground between the trailer and the towing vehicle. AWG #10 or #8 wire is preferable for this type of installation, except where a specially small size is needed to balance the system when a single diode is used.

However, these large wires do not fit in standard trailer connectors. A heavy-duty portable cord connector or the polarized unit shown in the lower-right portion in Fig. 4-1 could be used. A good individual plug and jack type connector, rated at 25 amperes is shown upper-right. These are the Superior Electric Company RP25GR (red), or RP25GB (black) jack and the PS25GR (red), or PS25GVB (black) plug. Note that the connectors in both units are protected from accidental shorting. Another plug and jack, called the jumbo banana, shown lower-left portion can carry the current, but the tip is not protected. These are suitable for grounding leads, but should not be used for a hot connection. The part numbers are the E. F. Johnson #770 plug and #760 jack. The plug will also fit into an AWG #6 wire solder terminal, shown far left. This terminal may be much easier to mount than the regular jack shown.

At the top-left corner are shown aircraft connectors that make good water-resistant trailer connectors. A part number that will handle two AWG #8 wires is the MS3102-20-23S receptacle and the MS3106-20-23P plug. They are available as surplus or from electronic supply companies. One that can handle two AWG #8's and seven AWG #16 wires is the MS3102-20-16S receptacle and MS3106-20-16P plug.

TWIN-ENGINE SYSTEMS

Twin-engine installations can use any of the systems described if separate auxiliary batteries are used for each engine. However, the same (or one) auxiliary battery cannot be used with two engines. To attempt to do so involves paralleling generators or alternators which can operate at different speeds: don't try it, except for Class 4 generating systems. Even then the results will not be satisfactory.

When two Class 4 generating systems, as are commonly used in outboards, are connected in parallel, the engine with the high output takes over most of the charging load, and the other engine contributes little. The available output of the second engine is therefore lost. The connection of Fig. 3-7 will prevent this loss. Both engines contribute to the electrical system in accordance with demand. Each system operates as the manufacturers intended; there is no modification of the basic charging circuits. Yet there are two batteries, one reserved for starting, and one used for the cabin loads only.

If one engine slows or stops, the diodes in the alternators protect the slowed engine from reverse-current flow from the other engine using the battery. In the case of DC generators, however, there would be no such protection, so the slowed unit would attempt to run as a motor; it could draw sufficient

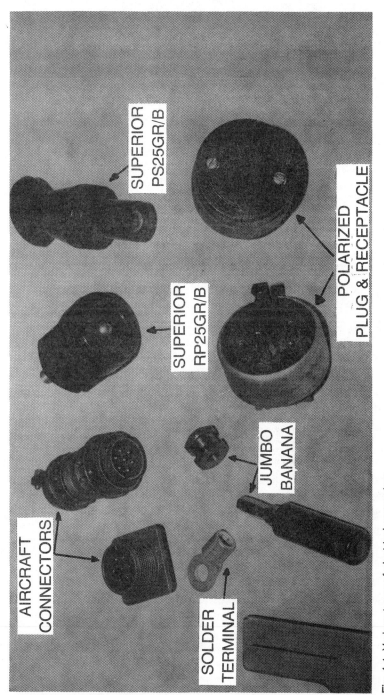

Fig. 4-1. Various types of electrical connectors.

current to destroy itself. At a certain point, the magnetism of the field could be reversed, compounding the problem. Never attempt to operate DC generators in parallel when they are driven by separate engines, or even by separate belts.

When an attempt is made to parallel two automotive alternators with regulators of any type (regulators can never be set exactly alike), the unit with the lower voltage setting will respond first to the system voltage. It will cut off its associated alternator and drop the load. However, if the load is quite high, the voltage of the second alternator may be pulled down sufficiently to cause the first one to attempt to pick up the load. The cycle then repeats, and the system oscillates or hunts.

If only a single auxiliary battery must be used, then the hookup of Fig. 3-7 can be used. However, it must be realized that the generating capacity of the engine connected directly to the starting battery is not being used to its fullest extent. This engine will charge up the starting battery and then idle along, while the other engine is required to carry all of the cabin loads—often a substantial amount in a large boat. To split the load evenly, two auxiliary batteries and a split-load system are required.

Another paralleling combination problem can occur if an attempt is made to charge a battery with a propulsion engine and an auxiliary power plant (APU) at the same time. Since the engine would normally have a regulated alternator, but the APU would not be regulated, it is difficult to predict what would happen. If the APU is equipped with a manual voltage regulator and an ammeter, it may be possible to obtain an adjustment where the APU carries some load some of the time, but it may be difficult to obtain a stable system. In a boat, where the propulsion engine runs at a constant speed for long periods, such a system may work, but in a land vehicle, where the speed is constantly changing, a satisfactory operation is doubtful.

The connections for *Method A*, are simple. Parallel the two batteries by connecting both positive terminals and negative terminals together, using a wire size equal to or better than the gauge used for the vehicle's starter cables. Since there is not enough room on the battery posts to attach two cables you will have to make a Y-connection for the ground and hot leads. You can do this easily by using two short cables on the positive terminals and two more on the negative terminals. Then bolt the two positive cables together. Do likewise with the negative cables. This works much better than trying to splice the cables.

With manual switch and contactor hookups, Methods B and C are also easy. Connect one terminal of the switch or contactor to the output terminal of the alternator or generator, leaving all existing vehicle wiring in place. Connect the other end of the switch or contactor to the wire leading to the second battery. If the contactor coil is not already grounded to its frame, connect one coil lead to ground and run the other to the control switch (ignition).

Some starter contactors have two terminals that look like coil terminals but one is used to short the ignition coil resistor. If it is marked, there will be the letter *S* near it. If it is unmarked, determine proper terminal by connecting a battery between the frame and a terminal. The solenoid will click when the proper terminal is touched.

Some care is necessary in connecting diodes for Methods D1 and D2. Proper polarity must be observed. In negative ground systems, the diode must be connected so that the arrow on the diode faces away from the alternator as shown in Fig. 4-2. The reverse is true in positive ground systems.

The diode for the vehicle system must be connected between the alternator's output terminal and *all* of the vehicle

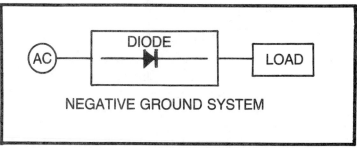

Fig. 4-2. Proper method of connecting a diode in a charging circuit. Connect the side of the diode with the arrow to the source in negative ground systems. In a positive ground system you must reverse the polarity of the diode. As a general rule in negative ground systems, point the arrow in the direction that you want the power to go.

loads. All wires on the alternator terminal must be removed to place the diode in series with them and the alternator.

The voltage-sensing lead of the regulator, if it is on the alternator terminal, must come off also so that it can be placed on the battery side of the diode. The charging lead to the secondary battery for Method D1 must go directly on the alternator's output terminal. Also, the lead into the second diode for Method D2 must go on this terminal. If you bought a dual diode isolator, the interconnection of the inputs of the two diodes will already have been made, so only one lead will go to the alternator's output. The secondary-battery charging wire goes to the output of the second diode.

Some fabricated isolators cannot be used on positive ground systems because the polarity of the diodes have to be reversed, and the connections are not accessible.

When checking the connections of diodes, do not go by the shape of the body of an unmarked unit. Diodes are made to conduct in only one direction, so a polarity check with a battery and voltmeter should be used. The procedure is to connect one end of the diode to the positive terminal of the battery. Use a short jumper wire for this purpose. Then connect the voltmeter's positive lead to the other end of the diode. Connect the common lead of the voltmeter to ground. An indication on the voltmeter will be noted if the diode is connected properly. It should read approximately battery voltage. If it is reversed there will be little or no indication on the voltmeter. Of course, this procedure is for negative ground systems. When testing the same diode in a positive ground system the leads of the voltmeter will have to be reversed, and the diode is connected to the negative battery terminal. Be careful of using an ohmmeter, because some of the lower cost ones have a different polarity at the leads when checking voltage or ohms; the common (black) lead may actually be positive when the meter is being used as a ohmmeter. There is no question with a battery and a voltmeter.

No harm will result if the diode is connected backwards between the alternator and the battery. There will simply be no flow of charging current. However, if a diode is grounded or connected across a battery in the forward direction, it will be destroyed. The safest precaution to use if there is any ques-

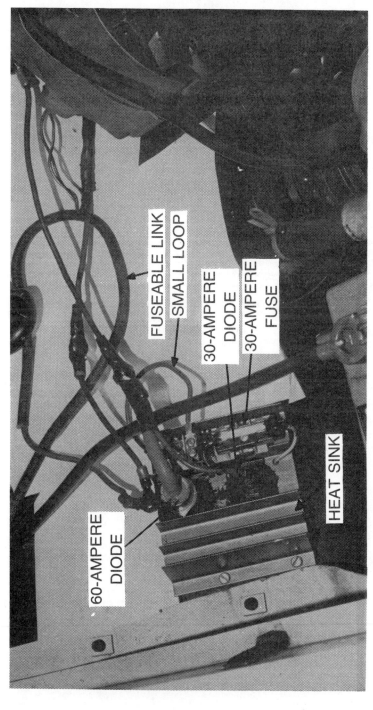

Fig. 4-3. A homemade diode isolator in a GM car. Basically, this is the same hookup as in Fig. 3-6 with some protective fuses.

tion about a connection is to use a fuse in series, or a fairly large light bulb of the correct voltage. Use a large one so there is no confusion of a wrong connection of just leakage. Leakage might cause a small bulb to glow somewhat.

Figure 4-3 shows the installation of a homemade dual diode isolator in a General Motors towing car. The two diodes are mounted in a heat sink (dissipator), bolted to the fender well. The large 60-ampere diode goes to the vehicle loads through the large looped wire that is actually a fuseable link (fuse). The trailer diode is a 30-ampere unit, connected to a 30-ampere fuse on top of the heat dissipator. The small loop of wire leads from this fuse to the trailer battery. The power output of the alternator goes directly to the diodes, and the voltage regulator wire that was also on the alternator terminal was moved along with the fuseable link to the output on the body of the large diode.

DUAL BATTERY RECOMMENDATIONS

- Method A is not particularly recommended since there is little to be gained except additional battery capacity. There is no way to protect the starter battery from being run down.
- Method D is not recommended for any 6-volt systems. The voltate drop across the diodes is too great of a percentage of the total voltage. Use a switch or contactor, Method B or C. See Fig. 3-1.
- Method D is not recommended for 12-volt DC generator systems unless you are willing to reset the regulator. Method B or C is recommended. See Fig. 13-10.
- Method D is not recommended for Class 3 alternator systems unless you are willing to reset the regulator. However, this cannot be done on some models. Method B or C is recommended. See Fig. 3-1.
- Class 2 systems can use any method of dual battery hookup. If there is a long charging lead to the trailer, Method D1 is recommended. See Figs. 3-3 or 3-4.
- Class 4 systems can use any method of battery hookup, but if the charging current is less than 10 amperes, Method E is recommended. See Fig. 3-7.

Chapter 5
Electrical Terms You Should Know

This chapter will attempt to explain enough about electrical terms, equipment, and principles to enable anyone to design his RV electrical system along the lines given in this book. No attempt will be made to explain electricity or why electrical devices work as they do. We are only interested in what they do and how to make them do it. If you already know electricity, you probably can skip this chapter. If you got lost in what has been said up to now, then read this chapter well.

Electricity acts a lot like a fluid—a supertransparent water that flows in solid metals (wires) but is contained by an insulator.

The pressure that causes electricity to flow is *electromotive force* (EMF), and it is measured in *volts*. This pressure is also called *potential*.

The flow of electricity is *current*, and it is measured in *amperes*. The flow of electricity in wires is impeded by a property called *resistance* that is measured in *ohms*. Resistance causes a loss in voltage along a wire. This loss is called *potential* or *voltage drop*. It is similar to the loss of pressure of a fluid flowing in a pipe.

Electricity can be stored in a tank, called a *capacitor*, whose size is measured in *farads*. Electricity can be stored in a magnetic field around a coil of wire, particularly if it is wrapped around an iron core. This device is called an *inductor*, whose size is measured in *henrys*.

Electricity can be converted and stored in a *battery* as chemical energy.

The amount of electricity that can be stored is measured by *watt-hours* or *watt-seconds*. When measuring storage or comparing devices at a constant voltage, the energy can be expressed as *ampere-hours* or *ampere-seconds*, but this is not a true measure of the energy.

Electrical pressure or *voltage* can be produced many ways: chemical action, mechanical forces, magnetism, heat, friction, etc. But, electrical *current* can only flow when voltage is impressed on the ends of an electrical *conductor*. Most of these processes can be reversed. Electricity can produce heat, mechanical forces, magnetism, chemical energy.

A *magnetic field* is produced around a wire or through a coil when electricity is flowing in it. But, if the current flow is stopped, the magnetic field collapses or disappears. There is energy in the field, so when it collapses it returns this energy to the electric circuit. This causes a spark, which you may have noticed when a switch is opened with an inductor in the circuit. The changing magnetic field *induces* an electrical voltage in the wire, and if the circuit is completed, current will flow. But this can happen only if the magnetic field is changing relative to the wire. It may be increasing or decreasing, the wire may move relative to the field, or the field relative to the wire. A steady magnetic field cannot induce voltage in a wire, no matter how strong it is. There must be a change taking place.

A voltage is generated by chemical action in a *battery*. A battery consists of several chemical *cells*. Each cell consists of two *electrodes*, usually of metal and a liquid, the *electrolyte*, which reacts chemically with the electrodes to produce the electricity. The generated electricity must flow internally through the electrolyte and out of the electrodes to the *terminals* of the cell. The voltage of the cell is determined by the composition of the electrodes and of the electrolyte. Cells are connected together in *series*, or end to end, to increase the voltage of a battery. Cells are connected together in parallel to increase the current capacity of a battery, but then the voltage stays the same as that of a single cell. The series connection increases the voltage, but the current is the same through all the cells.

The terminals of the cells (and battery) are considered to be *positive* or *negative*. Electric current flows from the positive to the negative terminals. The flow direction was selected by Benjamin Franklin when he was experimenting with lightning and electricity. Unfortunately, he made the wrong choice; the true flow of electricity is the other way. However, unless you are studying electronics, Ben's convention is used.

When a capacitor has electricity stored in it, there is voltage apparent on its terminals just like in a battery. So, when the circuit is closed, current will flow. But the time it flows is very short, because the amount of electricity in a capacitor is very limited. It is like a small water tank that is soon emptied by a big hole.

In an *inductor,* however, while electrical energy is stored in the magnetic field, there is no voltage apparent on the ends of the wire or coil. It is only when the magnetic field is collapsing that there is a voltage. At that time, completing the circuit will cause a current to flow. As soon as the magnetic field has disappeared, the current ceases.

A magnetic field can also be produced by a *permanent magnet.* The earth is a *permanent magnet.* Magnetic fields can attract or repel each other with a physical force. Magnetic fields attract pieces of iron or coils and wires having electricity flowing in them.

The flow of electricity can be steady as is produced by a battery connected to a circuit. This is called *direct current* or *DC*. The older generators of autos produced direct current by converting mechanical rotary motion of a shaft into electrical energy. This energy was stored in a battery by converting it to chemical energy.

Electrical flow can also be pulsating. The flow from an electrically charged capacitor is a pulse that lasts for only a short time. The current from a collapsing magnetic field is a pulse. Pulses can be made also by interrupting a direct current with a switch driven by a rotating cam as is done in the distributor of your car. If the pulses repeat regularly, the number of times per second the pulse repeats is measured by *hertz* (not the car leasing company!). One hertz is one pulse per second.

A pulse goes from zero to some value and then goes back to zero. The polarity of the pulse may be positive or negative, but not both. If the electricity goes from zero to some positive value, returns to zero, goes to some negative value, and again returns to zero, the current is said to be *alternating current*, or *AC*. If the cycle repeats, the *frequency* of the alternation is also measured in hertz. The time for an alternation to go from zero to plus, to zero to minus, then to zero is called the *period*, measured in *seconds*.

In the newer vehicles, an *alternator* produces AC, but this must be *rectified* by devices called *diodes* into DC, which is the type of electricity needed to charge a battery. Pulsating DC (never swinging beyond zero) will also charge a battery because it is always flowing in the same direction, albeit by fits and starts.

The distributor cam operates a switch (ignition points) to chop the DC from the battery into pulses that flow through the ignition coil. Each pulse of current generates a magnetic field in the coil. When the current is cut off by the points, the magnetic field collapses. Since the field repeatedly builds up and collapses, its motion *induces* a voltage into another (second) winding in the coil. As the magnetic field is building up, the voltage induced in the second winding has a certain polarity. But as the field collapses, the polarity that is induced is reverse. Therefore, the voltage induced in the secondary winding is alternating; it goes from zero to one polarity, to zero to the opposite polarity, to zero again, and repeats. This switch-and-coil combination is one way of *inverting* DC to AC. This combination does just the opposite of the *diode* or *rectifier* that converts AC to DC.

When two coils are wound on the same core, the voltages are *transformed* or charged in value if the number of turns in the two coils are different. Only AC and pulsating DC can be transformed. Smooth DC cannot because it does not cause the magnetic field to change. Since the secondary coil of the ignition coil has about a thousand times as many turns as the primary coil where the chopped DC is flowing, the output is transformed to about 12,000 volts. That is, 12 volts times 1000 equals 12,000.

DC cannot flow through a capacitor. AC alternately charges and discharges a capacitor each cycle, so in effect it flows through it.

A *diode* acts as a one-way check valve for electricity. It will pass DC in one direction and block it in the other. If AC is applied to it, it will pass the pulses of one polarity and block the opposite ones. The resulting current flow is *pulsating* DC. Current flows for half of the AC period, and not for the other half, so this is called *half-wave* rectification. It is possible to connect two diodes to a coil so that two pulses of current can flow during one period. This is called *full-wave* rectification. If there is only one coil involved in producing the AC, the system is called *single phase*. If two coils are used, the system is *two phase*. If three coils are used, the system is *three phase*. Normally, each coil has to have two wires, but 3-phase systems can be connected together in a Y-configuration or in a delta configuration. Now only three wires are needed instead of six to carry the power of the three coils. It is also possible to attach a wire to the junction of the three coils in the Y-connection. This is called a 4-wire, 3-phase system. (See Fig. 5-1.)

Some power will be transmitted in a 3-phase system even if one wire or coil is open.

Six diodes can be connected to the 3-phase system so the rectified output of all are added together. The pulse frequency resulting is three times as high as a 1-phase system; the output is much smoother.

Automotive alternators are made three phase to take advantage of all these factors, plus the fact that much less iron and copper is needed for a given output over a 1-phase alternator or even a DC generator.

The voltage induced in the windings of the older DC generators is actually alternating, but it is converted to DC by the brushes and commutator of the machine. These parts are *mechanical rectifier* that does the same thing as the diodes in the modern alternator.

An *insulator* normally will not allow electricity to pass through it. Air, gases, plastics, dry wood, oils, and most nonmetals are insulators. Many insulators will conduct electricity if they get wet, or are heated very hot. Also, if the voltage is raised enough, it will break the insulation down, and there will be conduction of current. The flash of a spark plug is an example of such a high-voltage breakdown.

When electricity flows in a circuit, there is energy moving, and the rate of movement is *power*. The current can be either in one direction (DC), or it may be back and forth (AC). This current and the voltage driving it act together to produce power. The measure of power is the product of volts and amperes, and it is called *watts*.

Mechanical power is measured in *horsepower*. There are 746 watts in 1 HP (horsepower). Power can be measured by a *wattmeter*, or it may be calculated as the product of the volts and amperes in the circuit.

Electrical power is utilized by being converted into other forms of power or energy, such as heat in a resistor, light in a lamp, mechanical power in a motor, force in a solenoid, chemical energy in a battery. The conversion is never perfect, however; there is a loss of some power every time there is a change. The loss is measured by the *efficiency* of the converter. There is even a loss of power when current flows in a wire; the *resistance* of the wire converts the electricity to heat as it flows along.

Sometimes quite complex devices are designed to minimize the losses in conversion of power. An incandescent lamp is a very inefficient converter of electricity into light. A much more efficient device is the fluorescent lamp. But to operate a fluorescent lamp on 12 volts DC, it is necessary to convert the DC to AC, transform the voltage to over 100 volts, and control the current through the lamp with an inductor. The result is worthwhile because more light can be obtained for 2 amperes in a fluorescent lamp than is available from 6 amperes in an incandescent light. Also, there is considerably less heat given off.

MEASURING ELECTRICITY

Electricity is invisible, so one cannot estimate how much is flowing in a wire as is possible with water flowing in a pipe or

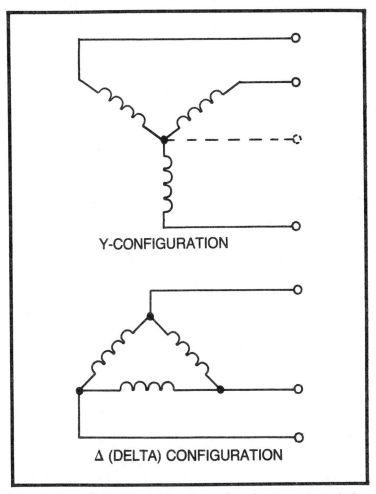

Fig. 5-1. Two methods of connecting the windings of 3-phase motors. In the Y-configuration, a fourth wire can be attached to the center junction to obtain a 4-wire, 3-phase system. The delta configuration resembles the Greek letter, thus its name.

ditch. So, it must be measured with instruments or meters. When such measurements are made, the relationship with the other factors in the circuit can be found by simple mathematical formulas. They are not difficult to use and with the data given in this book, will allow you to solve most of the common RV electrical problems.

Ohm's law is the basic relationship between voltage, current, and resistance.

E = voltage = Electromotive force or pressure
I = amperes = Electrical current or flow
R = ohms = resistance of a wire or part

$$E = I \times R$$
$$I = E/R$$
$$R = E/I$$

Power formula:

P = Watts = electrical power = E × I
746 Watts = 1 HP

$$E = W/I$$
$$I = W/E$$

By combining the two formulas, the following are obtained:

1) $P = E^2/R$
2) $E = \sqrt{P \times R}$
3) $R = E^2/P$
4) $P = I^2 \times R$
5) $I \times \sqrt{P/R}$
6) $R\, P/I^2$

EXAMPLES

If you are not used to working with formulas, the simplest way to use them is to find the proper one first by making a table of all the *known and unknown factors,* setting each equal to the letter that it represents and setting question marks after the unknown letters. Then, with this table, go through the formulas until you find the one with all the known factors. If you are unable to find one with sufficient known factors, then more measurements are required, or you have a problem not covered by the formulas given.

Example 1

According to the wire tables, *Chapter 8*, a 20-foot piece of AWG #14 wire has a resistance of 0.05 ohms. With a current of 10 amperes flowing, what is the voltage drop in the wire?

Known   R = .05 ohms
        I = 10 amperes
Unknown E = ?

The proper formula relating the quantities is:
$$E = I \times R = 0.05 \times 10 = 0.5 \text{ volts}$$

Example 2

A light bulb is marked 12 volts, 40 watts. What is the current drawn by the bulb?

Known $P = 40$ watts
               $E = 12$ volts
Unknown $I = ?$

The proper formula is:
$$I = P/E = 40/12 = 3.33 \text{ amperes}$$

Example 3

By cut and try, it is found that the right size of series resistor for proper action of an electric trailer brake set has 2 volts drop across it when the brakes draw 4 amperes. In order to buy a resistor for this application, the resistance and the power rating (watts) of the new part are needed.

Known $I = 4$ amperes
                 $E = 2$ volts
Unknown $R = ?$
                 $P = ?$

The formula for resistance is:
$$R = E/I = 2/4 = 0.5 \text{ ohms}$$

The formula for wattage (power) is:
$$P = E \times I = 4 \times 2 = 8 \text{ watts}$$

Eight watts is not a standard rating for a resistor, so the next larger size, a 10- or 12-watt unit can be used. The larger size will handle more power, so it will actually run cooler in this application.

This problem points up another limitation of resistors. The wattage rating is based on current flowing through the entire resistor, and a portion of it, half for instance, can only dissipate half of the total power. What this means is that there is a maximum current limit that cannot be exceeded in any portion of the resistor. The limit is given by the power formula:

$$I = \sqrt{P/R}$$

In the chapter on trailer brakes, where such large resistors would be used most often, is given a table of resistors sizes, ratings and current limitations.

ELECTRICAL MEASUREMENTS

Voltage is the electrical pressure or force between two wires or terminals. It is measured with a voltmeter, having test leads which are applied across the two points.

The voltmeter actually draws some current in order to indicate. The amount of this current is expressed indirectly in the specifications for the meter by giving the *ohms-per-volt* resistance. Typical values may be 100, 1000, 20,000 ohms-per-volt. In general, the lower the number, the more rugged the meter. Values above 1000 ohms-per-volt are only needed to test electronic circuits. But 100- or 1000-ohms-per-volt meters are quite satisfactory for automotive and RV testing.

Electrical currents are measured by an *ammeter* that is normally inserted in series in the wire where the current is to be measured. An ammeter has a very low resistance, usually less than 0.01 ohm; so, it is fatal to the instrument to connect it *across* two terminals of a battery or generator as is done with a voltmeter. In case of doubt, use a voltmeter first across the terminals; if there is no voltage reading, it is safe to connect the ammeter. There is an ammeter designed to test battery output that can be connected directly across the battery, but it is specially designed for the job.

There is another type of ammeter now available that merely has to be laid across a wire to give a reading. The reading is fairly good provided the wire is not near a piece of iron or another wire (1 inch or more away). This type of meter cannot be used in a bundle of wires, or where two wires next to each other are carrying the current to and from the load, as in a twin molded lamp cord. Such a meter will read zero on that twin cord no matter what the current. The meter depends for its operation on the magnetic field around a wire, but the current going one direction will neutralize the magnetic field of the current going the other direction along parallel wires. This is a good idea to prevent deflection of a compass in a boat, but not so for this type of ammeter to read.

Ammeters are of three basic types: moving magnet, self-contained moving coil, and externally shunted moving coil. All do the same job with some differences in accuracy and a big difference in price. The moving-magnet type, as made by Eimco, Hoyt and Shurite are adequate for RV use.

Ohmmeters measure the resistance of a part or wire, but unfortunately, the most readily available instruments do not have the ranges really required for RV and automotive use. Most small electronic meters have ranges too high, measuring easily hundreds of thousands or millions of ohms, whereas the vehicle user needs a meter that will measure tenths and hundredths of ohms. The midrange scales of these meters, from 100 to 1000 ohms are generally useful for continuity checks, for measuring relay coils, light bulbs, or similar devices. Be careful never to place an ohmmeter across a live circuit, as the high voltage of the circuit will burn out the meter. Most ohmmeters are designed to work on one 1.5-volt cell.

When an extremely low resistance has to be measured it must be arrived at indirectly by sending a fairly large current through the unknown, then measuring the current and voltage drop. The resistance is then calculated by Ohm's law.

Electrical quantities are related by multiples of ten. Most of the generally used combinations have been given names:

Giga	10^9	1,000,000,000	(Billion)
Mega	10^6	1,000,000	(Million)
Kilo	10^3	1000	(Thousand)
Hecto	10^2	100	(Hundred)
Deka	10^1	10	(Ten)
Deci	10^{-1}	1/10	(Tenth)
Centi	10^{-2}	1/100	(Hundredth)
Milli	10^{-3}	1/1000	(Thousandth)
Micro	10^{-6}	1/1,000,000	(Millionth)
Pico	10^{-12}		(Thousand Billionth) (Million Millionth)

For example:

1 kiloampere =1,000 amperes
1 milliampere = One thousandth of an ampere
1 picoampere = One million millionth of an ampere
1 megohm = 1 million ohms

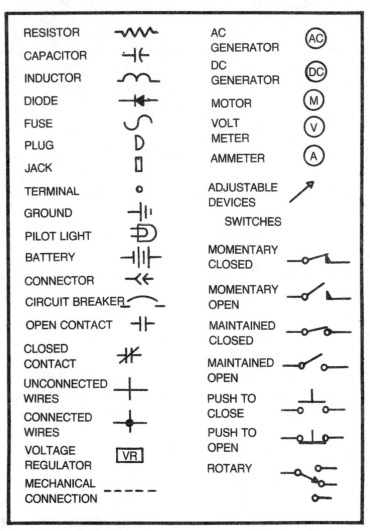

Fig. 5-2. Electrical symbols. These are some of the more frequent ones that you will see in RV applications. There are many more and many variations of the ones shown.

1 kilohm = 1,000 ohms
1 milliohm = One thousandth of an ohm
1 microhm = 1 millionth of an ohm

ELECTRICAL WIRING DIAGRAMS AND SCHEMATICS

Electrical schematics are an electrical short-hand showing how to interconnect electrical parts and devices. They do not show the exact routing of the wire between the parts. The wiring diagram shows detail routing and terminal connections. The schematic is drawn more for the convenience of the draftsman or the person studying how a circuit works. It is also very useful for making one-of-a-kind projects because it shows functional interconnections: as long as two parts function the same, they can be connected the same regardless of the actual terminal numbering or positions. But the person doing the wiring must understand the symbols used in the diagram and how they apply to the device to be used.

Production wiring diagrams, as typically given in automotive manuals do not explain functions. If the new part does not have exactly the same terminals, it cannot be connected properly.

Since most of the projects in this book are one time, or experimental, schematics are used. Figure 5-2 gives the more common symbols used for these diagrams.

Some discussion of the symbols for relays and contactors is needed since there are two widely different methods of showing them by electronic people and electrical power workers.

Electronic relays are generally classed as low power, and are most often used for control, sequencing, and logic functions. The contacts are shown by symbols like momentary switches, and the coils are shown by the inductor symbol. The coil and all of its associated contacts are shown grouped close

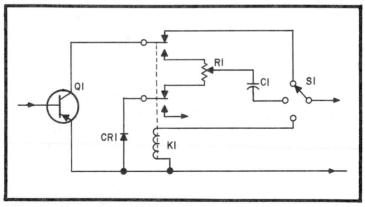

Fig. 5-3. Method of illustrating a relay and its contacts in an electronic schematic.

together with the wires routed as necessary to this group symbol as is shown in Fig. 5-3.

Electrical power uses deal with two classes of relays. The first is similar to the electronic in function: logic and sequencing. The contact loads are generally low. The second class of units are power contactors where the contacts handle high-power circuits. Normally, the schematic does not differentiate between the two. The contact symbols are two short parallel lines, with normally closed contacts indicated by a slash across the two.

Coils are usually designated by circles, and when there are many relays involved, by a designating letter or number within the circle. The contact symbols are positioned anywhere on the drawing, as dictated by the wire runs. Contacts and their operating coils are almost never together. Contacts may be keyed back to their coils by coding with the coil designator and another letter or number if there is more than one contact per coil. For example, K1A and K1B contacts are operated by coil K1. The electrical method of showing relay coils and contacts is shown in Fig. 5-4.

There is no sharp distinction between what is called a relay or a contactor. A generally accepted practice is to call any logic or control unit a relay, at least as long as the contacts are handling less than 15 or 20 amperes. A unit designed for handling larger currents or high voltages is called a contactor.

A relay or contactor has two separate ratings, that of the coil and that of the contacts. The coil is usually rated by the proper voltage to operate the relay normally. This may be DC or AC, but not both. The coil resistance, wattage, or current draw may also be given. For AC coils, the *volt-amperes* value is given rather than watts.

A coil will pull in (energize) at a somewhat lower voltage than it is rated, but a point is finally reached where it will not operate. However, once energized it may hold it to a very low voltage. This *dropout* voltage is seldom controlled by the manufacturer, so it can vary widely. The *pull-in* voltage is fairly closely controlled.

The contact rating of a relay or contactor varies widely with the type of current being handled, the properties of the load and other factors. Often several ratings like 100 amperes,

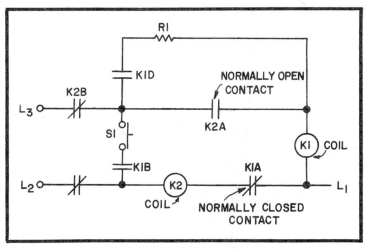

Fig. 5-4. Method of illustrating relays in a electrical power diagram.

24 volts DC, noninductive load, 20 amperes 115 volts AC, 10 amperes 230 volts AC; or other combinations will be given. There is no relation between coil and contact ratings.

As a guide, for 12 volts, any 12-, 24-, or 32-volt DC rating can be used as is. A 120 volt AC rating (or current) can be doubled, but there is no way of relating to 120 volts DC to 220 volts AC.

The type of contact operation of switches and relays is described several different ways. The older way was to say single-pole, double-throw action or similar. A common way for military units is to call this *form C*. A table of the different designations is given here:

RELAY AND SWITCH CONTACTS

FORM	NAME	DESCRIPTION
A	Single-pole, normally open	Single circuit, shown in open position. Closed when operated.
B	Single-pole, normally closed	Single circuit, shown in closed position. Opened when operated.
C	Single-pole, double-throw	Single circuit with one open and one closed contact. Arm transfers when operated.
2A	Double-pole, normally open	Two circuits as above in A.

55

FORM	NAME	DESCRIPTION
3A	Triple-pole, normally open	Three circuits as above in A.
2B	Double-pole, normally closed	Two circuits as in B.
2C	Double-pole, double-throw	Two circuits as in C.

Some relays are made with a double series-break arrangement rather than a swinging contact. Also, some control relays are made with break-before-make or make-before-break contacts sequences. These are special types that normally would not be used in vehicles.

On the other hand, power contactors are often made with double series-break contacts. Both contactors in Fig. 3-2 are of this type. For all practical purposes, these operate like form A, B, or C contacts as described above.

Chapter 6
Battery Basics

There are two basic kinds of batteries. Both operate on the same principle of converting chemical energy into electrical energy.

Primary batteries are not rechargeable. They are used once and discarded. Flashlight batteries are typical primary batteries (Fig. 6-1).

Secondary batteries are rechargeable with electricity from some other source such as a generator. The electricity in the battery can be used up, then the battery is connected to a source of electricity and *recharged*; a new quantity of electricity is stored in it. The source of the new charge can be another battery, a generator, or any other source of *DC*. Special devices which convert *AC* such as are available from a house plus, into DC, especially for charging batteries are called *chargers*. They have special circuitry to regulate the amount of the charge to prevent damage to the battery (Fig. 6-2).

CONSTRUCTION

Strictly speaking, a *battery* is a number of *cells* connected together as desired to produce a given voltage or current output. A *cell* (Fig. 6-3) is the minimum combination of parts required to generate electricity by chemical activity. Cells are

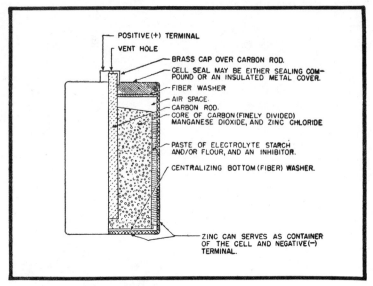

Fig. 6-1. A primary or dry cell.

connected in series (end to end) to produce higher voltage batteries, and in parallel to produce higher current batteries.

There are many ways to make cells and batteries, but only one will be considered here. It is the commonly used *lead-acid* battery. It consists of three or six cells connected in series. Since the nominal voltage of each cell is about 2 volts, the batteries are *rated* at 6 and 12 volts.

In each cell is a set of *positive plates* or *electrodes*, and a set of *negative plates* or *electrodes*. The positive and negative plates are separated from each other by porous insulators. The cell is filled with a liquid that is a mixture of sulfuric acid and water. This liquid is called the *electrolyte*. The chemical reaction of the acid with the lead plates generates the electricity. As the electricity is used, the acid reacts with the lead and leaves the water. Since the acid is much heavier than the water, the mixture becomes lighter as it loses acid. So, the weight of the electrolyte is a measure of the electricity used.

MEASURING BATTERY CONDITON

The weight of the electrolyte is called *specific gravity*, and it is measured with an instrument called a *hydrometer* (Fig. 6-4). Temperature also affects specific gravity, so the better

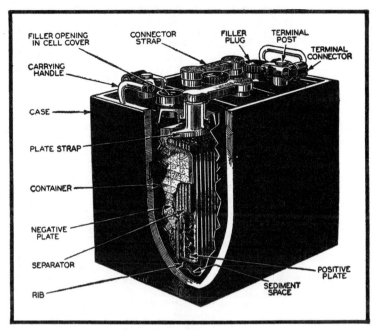

Fig. 6-2. A cutaway of a 6-volt secondary battery.

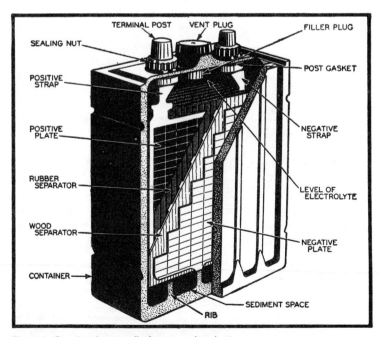

Fig. 6-3. One 2-volt wet cell of a secondary battery.

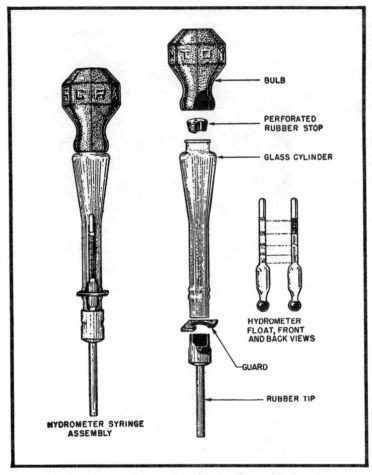

Fig. 6-4. Hydrometers used to measure the specific gravity of a battery.

hydrometers are made with a thermometer that shows the correction that needs to be made if the temperature is too far from the normal 80° Fahrenheit (Fig. 6-5).

Specific gravity is actually a ratio of the weight of the acid mixture to pure water. For batteries it is usually expressed as *points*. A reading of 1250 points, for instance, is actually 1.250 specific gravity. The mixture is 1.250 times as heavy as water. The reading on the hydrometer stem is in points, and a temperature correction is given in points.

Typically, a battery used in a colder climate (northern USA) will have a hydrometer reading of 1275 points when it is fully charged, going down to about 1100 points when it is discharged. For a warm climate (southern USA) the fully charged reading may never exceed 1250, and the discharged reading can go below 1050.

A good battery will normally have less than 50 points of specific gravity difference between any of its cells. If the difference exceeds this value, the low cell may be going bad. Another possibility is that the battery is not being charged enough. To check this, the battery should be overcharged

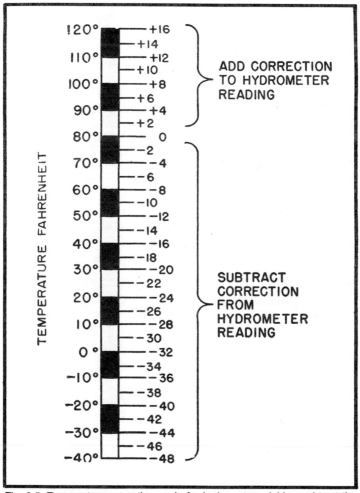

Fig. 6-5. Temperature correction scale for hydrometers. Add or subtract the amount of points shown on the right-hand scale according to the temperature.

until all the cells are gassing freely for several hours. Then recheck the gravity to see if it has come up. If after several days, the cell is low again, it is defective.

The actual voltage of a battery varies with the state of charge, but only rapidly near exhaustion of the charge. A fully charged 6-cell battery may read 12.5 to 13.2 volts; half discharged, 11.8 to 12.2 volts; three-quarters discharged, 11.7 to 11.3 volts; and fully discharged, less than 11 volts. However, when a load is applied, the voltage of a full battery drops only a small amount until it is half empty, then the voltage drops much more rapidly. Any car battery that shows a voltage of less than 10 volts when 12 or 15 amperes is placed on it is either completely discharged or defective.

BATTERY CHARGING

The action of a battery while it is being charged at 10 to 12 amperes is about like this: The acid gravity builds up fairly smoothly to within about 50 points of full charge with some gas being evolved. Then the gassing rapidly increases, and if the rate of charge is continued, the electrolyte is spattered out of the vents, the liquid level rapidly lowers and the battery gets hot. The gassing is actually caused by the breakdown of the water in the battery into its elements, hydrogen and oxygen. When heavy gassing starts, or the voltage of the battery reaches 14 volts, the charge rate should be cut down to about 3 to 4 amperes to prevent this loss of water and overheating.

Incidentally, spattered electrolyte around the vents of a battery in a vehicle is evidence of heavy gassing so the cause should be investigated before the life of the battery is shortened. The final electrolyte gravity and battery voltage is reached after several hours of charging at the lower rate. The amount of electricity put back in the battery is about 115% of the amount used.

The overvoltage of the charger above the battery voltage drives the charging current into the battery. Only a small change in the charger overvoltage can make a large difference in the charging current. Any excess of charging current (and quantity of electricity forced into the battery) over the capacity of the battery will be lost as gassing and heating.

If a battery is not used for a long time, it will gradually lose its charge. If a battery is allowed to stand for a long time and it does lose its charge, it will be very difficult to recharge. The slow loss causes a somewhat different chemical reaction in the cells, and the resulting chemical compound is not so easily converted as when the discharge is rapid. Also, a discharged battery, if allowed to stand for a long period of time will have the same thing happen. It is called *sulfation*. Heavy-duty industrial batteries can be partially recovered from sulfation, but the typical auto battery, once it is sulfated, might as well be thrown away. If a battery will not accept any charging current when 15 or 16 volts is applied to it, it is probably sulfated.

If you wish to try to salvage a sulfated battery, a charging device capable of applying up to about 20 volts may be needed. Apply whatever voltage is necessary to put 1 or 2 amperes into it. But watch it very carefully, for in an hour or two, if it is going to react at all, the current will start to go up and the voltage must be dropped. Do not run more than 3 to 5 amperes into it for about 10 hours. If the battery has picked up some charge, say to an electrolyte gravity of 1100 or more, discharge the battery with a twentieth of the ampere-hour rating current for 10 to 12 hours, then recharge at 5 amperes overnight (12 − 15 hours). Discharge it as before and recharge again, until all the cells are gassing freely. If the gravity of the cells goes above 1230, that is the best that can be done. The higher the reading, the better the salvage. Do not add acid to a sulfated battery to bring the gravity reading up; the excess acid will eat up the plate support grids.

To prevent sulfation, fully charge the battery before storage and give it a light charge of about 1 ampere for 8 hours or 0.5 ampere overnight every 3 or 4 months. Do it every 2 months if the storage temperature is over 80° Fahrenheit.

MAINTENANCE & REPAIR

Be sure the liquid level is well above the plates at all times as a low liquid level will cause corrosion of the plate support grids and the plate will break off. Minor corrosion erodes the grids and causes high internal resistance that limits the battery capacity.

Keep the top of the battery dry and free of dirt and acid. If the terminals corrode, clean them thoroughly and coat lightly with grease. Sometimes, a battery will leak acid around a post and the terminals cannot be kept clean. There are battery case repair materials available that will seal the joint, but if it is well cleaned and dried, roof patching asphalt will make a good repair.

The best method of cleaning up acid spills or spatters is by brushing a saturated solution of baking soda in water over the area. Do not get the material inside of the battery. Heavily corroded wires and terminals can be cleaned by soaking them in the same solution. Place the terminal in a cup or glass of water or the solution and sprinkle more soda over it. Allow it to soak with occasional stirring or brushing until all the corrosion is removed.

Exposed copper or brass will soon corrode again even if it is protected with grease. The only way to protect the part is to cover it with solder by dipping in a pot of molten solder or flowing solder over it with a torch. Such a repair is not as good as the original lead coating but it will do in an emergency.

WARNING: The mixture of gases given off by a charging battery is hydrogen and oxygen mixed in the proper proportions to cause a violent explosion. Always vent the box containing batteries, preferably at the top and bottom. Never use a switch or relay in the box that can cause a spark, or have the box near an open flame. A cubic foot of this gas mixture can destroy an entire camper if it explodes! Remember that this spark can occur if you pull off the charger leads from the battery before turning off the charger or disconnecting it from the power line.

BATTERY WATER

The best water to use in a battery is distilled or deionized water such as is sold in most stores for use in steam irons. The next best is rainwater collected only after the collecting area has been well cleaned and flushed, and that has been stored in nonmetallic containers.

Of the various harmful impurities, iron, chlorides, and nitrates are the worst. Water from rusty containers, old pipes, or that has mud in it should not be used. Chlorides are common in the West, and where they are used for melting ice. Common salt and road de-icers are chlorides. Nitrates are common in fertilizer or in rainwater that falls during a heavy lightning storm. The common hard water minerals, calcium and magnesium are somewhat harmful, so if your water is hard, do not use it in a battery. Never put antifreeze of any type, or any organic acid, like ecetic (vinegar) in a battery. The black manganese mixture from a dry cell is fatal. A battery will tolerate some copper, zinc, and silver, but the less the better.

BATTERY CAPACITY

Batteries are rated by voltage and ampere-hours. An ampere-hour is the product of amperes of current and the time the current flows in hours:

$$AH = I \times T = \text{amperes} \times \text{time}$$

The true electrical capacity of a battery is measured in *watt-hours*, which is *amperes* times *volts* times voltage, ampere-hours is generally used for convenience.

Voltage rating is determined by the number of cells in series. The nominal voltage of the wet cell is 2 volts, but the exact voltage at a given time depends on the temperature, the electrolyte concentration, and quite a bit on the current load on the battery and its internal construction.

A voltage reading with no current drain at all will not determine the condition of the battery. Even a badly discharged or worn out 6-cell battery will read 12 to 13 volts. But if a load is put on it, then the true condition is soon evident. If the gravity of the electrolyte is low and the voltage drops rapidly below about 11 volts when a load of about 10 amperes is applied, then the battery at least needs to be charged. But, if the gravity is high and the voltage drops rapidly, the battery is bad. It may have a leaky cell, the plates corroded from bad water, a low water level, or have had acid added when it is worn out.

Usually a battery fails first in one cell, and the trouble shows up when the gravity of that cell drops below the others. Sometimes, this low gravity is caused by not charging the battery fully, so a suspect battery should be charged for about a twentieth of the AH rate for 3 or 4 hours and then rechecked. For example, an 80 AH battery should be charged at 80/20 = 4 amperes for the specified time. If the gravity of the cell does not come up to that of the others, or if it drops still further after a period of 10 to 20 hours, the cell is bad.

The ampere-hour capacity of a battery is usually given at the *20-hour rate*. This rate is the amount of current flow that will just discharge the battery down to 1.75 volts per cell (10.5 volts for six cells), in exactly 20 hours. For example, if 4 amperes will just bring the cell down to 1.75 volts in 20 hours, the 20-hour rate is 4 amperes and the capacity is 4 × 20 = 80 ampere-hours.

The 20-hour rate is a current, but the *20-ampere-hour rating* is the capacity rating of the battery.

The ampere-hour rating of a battery actually varies with the current drain. As the current drain increases, the available ampere-hours decrease. This is the reason for some of the other ratings given for a battery. They concern the high-current drains for a short period of time involved in cranking engines, and are not usable for lighting or other long-time, low-current drains.

Below are listed the allowable current drains and available ampere-hour capacities of several standard automotive batteries when loaded for certain periods of time.

AUTOMOTIVE 24 SUPER

OPERATION TIME	CURRENT DRAIN (hours)	AMPERE-HOURS (amperes)
20	4.2	84
10	7.8	78
5	13.0	67
2.5	23.0	57

AUTOMOTIVE 24 STANDARD & 22 SUPER

OPERATION TIME (hours)	CURRENT DRAIN	AMPERE-HOURS (amperes)
20	2.5	50
10	4.6	46
5	8.0	40
2.5	13.0	34

AUTOMOTIVE 22 STANDARD

OPERATION TIME (hours)	CURRENT DRAIN	AMPERE-HOURS (amperes)
20	1.5	30
10	2.8	28
5	5.0	25
2.5	8.0	20

Six-volt batteries can be calculated the same way as far as current and available ampere-hours are concerned, but remember that two in a series will be required for a 12-volt system.

6-VOLT AUTOMOTIVE 1 SUPER

OPERATION TIME (hours)	CURRENT DRAIN	AMPERE-HOURS (amperes)
20	6.8	135
10	12.5	125
5	21.0	105
2.5	36.8	92

If you have a battery with a different ampere-hour rating than those shown above, the following percentages can be used to find the energy at the given times:

OPERATION TIME	PERCENT OF 20-HOUR RATING
20	100
10	92
5	80
2.5	68

To get the allowable current drain, divide the available ampere-hours by the time. It is not easy to work backwards—given a certain current drain and time of operation, to find the necessary battery size. One must guess at a size, calculate the rating and amperes; then try again if the answer is not what is required.

The proper charging rate for a low battery with the typical small home charger is about a twentieth of the ampere-hour rating of the battery: an 84 ampere-hour battery should be charged at 4.2 amperes (84/20). If your charger does not give this current, divide the ampere-hour rating of the battery by the actual charger current, and the answer will be the approximate time that the battery should be left on

67

charge. If the current drops at the end of this time to less than 1 ampere, it is safe to leave the charger on several more hours; but, if the current is between 1 and 3 amperes, the battery should be finished in an hour. If the current is over 3 amperes, there is probably something wrong with battery or the charger. If the battery is gassing and the gravity is high, the charger is suspect—it is putting out too high a voltage.

A battery can be charged at a much higher rate if the charger has a voltage regulator or timer to control the charge. For instance, this same battery can be charged at a tenth of the AH rating for about 8 hours, then the rate must be cut down to the twentieth rate or a little lower to finish. If this is not done, the battery gasses too heavily; it may be overheated and its life is shortened.

In the vehicle, the charging rate is regulated by the battery voltage and the temperature of the regulator. So, although the charge rate may start quite high, it soon drops down to a much lower value that will not damage the battery.

Many large industrial and service shop chargers also use the battery voltage and temperature to allow a faster charge without harming the battery.

The usefulness of a home battery charger can be considerably extended if it is fed with a variable voltage transformer as shown in Fig. 6-6. These transformers allow the output line voltage to be varied from zero to 120° of the input by a control knob on the top. When a charger is plugged into the output, the charge rate of a battery can be charged with a 12-volt unit if necessary.

The variable voltage units are available as surplus or from electronic mail order companies. To estimate the size unit needed, since they are rated in amperes, multiply the maximum charging current times the battery voltage, then divide by 120. This is a double application of the power law:

$$P = I_{CHARGE} \times V_{BATTERY}$$
$$I_{LINE} = P/120 \text{ (120-volt AC line)}$$

Pick the next larger current rating for the variable voltage transformer.

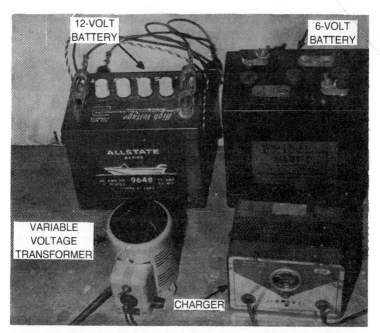

Fig. 6-6. Charging a battery with a charger connected through a variable voltage transformer.

These units are available in cases as shown, with meters, or in open, uncase types. Uncased units should be mounted in a box, since the exposed 120-volt terminals are a hazard.

ESTIMATING BATTERY LOADS

To go through the procedure of estimating the size of a battery again in detail:

1. Add up all the current drains that are on for long periods of time like lights, fans, radio, or TV. Multiply the amperes drain by the hours the load will be on.
2. Estimate the intermittent loads as best as possible, such as a water pump, toilet flush, radio transmitter, in terms of ampere-hours.
3. Add all the ampere-hour loads for both types.
4. Now using the *continuous* current drain only, look in the battery capacity table for a battery that will give the *total* ampere-hours needed for the time the *continuous* loads will be on.

Here: Assumes 2 days between charging.

For example the continuous drain is 4 amperes, 7 hours per day for 2 days, or a total of 14 hours, the ampere-hours are:

$$4 \times 14 = 56$$

Intermittent drain is 10 amperes for 2 hours, or 20 AH. It follows then that the total energy is:

$$56 + 20 = 76 \text{ AH}$$

What is needed is a battery that will operate for at least 14 hours at a continuous drain of 4 amperes, with a total available amount of energy equal to 76 AH.

The standard 24 size is too small, since it has only 45 AH at 4.6 amperes for 10 hours.

The super 24 size will give 84 AH at 4.2 amperes for up to 20 hours, so it is more than adequate. There are always some unrecognized drains, or new ones added, so it pays to go over the estimate as far as you can afford. Remember that the battery loses capacity as it gets older, too.

If no battery listed will handle the drain, it may be necessary to add two batteries in parallel, or two 6-volt ones in series, since they usually have a higher ampere-hour rating than the 12-volt type.

Where high capacity is needed, an economical choice may be golf-cart batteries. They have a 20-hour rating of about 170 to 180 ampere-hours, and they are designed for deep discharge at fairly low currents, so they hold better voltage near the end of discharge.

Anything larger than this will have to be selected from diesel or marine starting batteries. These batteries have two disadvantages: weight and cost. They are rugged, but it may be more economical to buy several automotive types and parallel them, particularly if the batteries are to be moved in and out of the vehicle often.

SYSTEM CHECKS

Having selected a battery that appears adequate, and with the charging system installed, how can everything be checked out?

Before you start, charge the new battery thoroughly and take the hydrometer readings for all the cells. Mark the values on the wall of the battery box where they will remain for the life of the battery.

If you have an ammeter or two, measure the charging current into both the starting battery and the secondary battery. Now, assuming the starting battery was not run down by a bulky engine start, both batteries should be charging at about the same rate. An exact correspondence is not necessary; but the values should be reasonably close together, with the starting battery taking more current.

After a thorough warmup, or a short road trip, the charging rate should still be fairly well balanced, but at a much lower level, say in the 2 − 5 ampere range. Again, an exact balance is not necessary.

Now use the loads in the cabin to run the secondary battery down, say to 75° charge as evidenced by a 60 − 100 point drop in the specific gravity.

When the battery is recharging with the cabin loads off, it should take from 50 − 80° more current than the vehicle starting battery for at least 3 or 4 hours. Then the currents to the two batteries should be more evenly balanced and of much lower value.

If all is well in the tests, the final proof of the system is a somewhat extended trip, where the batteries are used normally. Monitor the specific gravity of the RV battery. If it maintains an average over 1200, all is well. If the gravity drops, either some of the loads will have to be eliminated, the charging rate increased, or a larger battery used.

If the battery shows spattering of acid around the filler caps, or if the water level seems to decrease rapidly, the battery is being overcharged by too high a rate. If the gravity is high, there is no spattering, and the water level goes down rapidly, the rate is not too high, but the charging time is too long. Since this cannot be changed, the rate will have to be lowered to solve this problem.

The same comments apply to the vehicle starting battery. If both batteries are wet or show signs of gassing, the voltage regulator is probably set too high. If both batteries do not charge fully, the regulator is too low, or if a diode isolator is used with a Class 1 or 3 charging system, the voltage regulator is not measuring the actual battery voltage so the system cannot operate properly. If the voltage can be changed, increase it by the amount of the diode drop.

If too low batteries occur in the Method E, twin-engine system, the load is too great for the charging capabilities of the engines. The only answer is less load or an auxiliary power plant.

If the RV battery is gassing and has a high specific gravity, and the starter battery is normal or a little low, some resistance must be added to the RV charging line.

If the vehicle starting battery is gassing, and the RV battery is normal or low (a typical situation), a resistance will have to be added to the vehicle charging circuit.

Unfortunately, there is no commercial source for the necessary resistor. A suggestion is to make it out of about a foot of 0.5-inch (wide) by 0.012-inch (thick) stainless steel box strapping or similar material, cutting down the length as necessary to balance the system.

A voltmeter in the RV battery is an aid to monitoring its condition. One of the low cost, moving-magnet types with a 15 or 20 volt scale is quite good, and the price is about $6 – $8 from electronic mail order houses.

Remember that the voltage of the battery is high when it is charging and it is lower when loaded. The loaded voltage is the best measure of the battery condition. The danger signal is when it reads 11.25 volts when the load is about a twentieth of the 20 ampere-hour rating of the battery; 10.5 volts is the absolute lowest it should be allowed to go to.

It is important to check the original gravity reading of a well charged new battery because the gravity is not necessarily the same in batteries built for northern or southern climates. Also, the voltage varies with the gravity reading as the following list shows:

POINT-VOLTAGE READINGS, AUTOMOTIVE BATTERIES

CHARGE CONDITION	COLD AREAS		WARM AREAS		TROPIC AREAS	
	POINTS	VOLTS	POINTS	VOLTS	POINTS	VOLTS
100#	1280	12.76	1260	12.71	1225	12.56
75%	1230	12.56	1215	12.55	1180	12.37
50%	1180	12.37	1170	12.30	1135	11.93
25%	1080	11.71	1070	11.50	1045	11.40

When discharging a battery at the 5-hour rate, the voltage will drop another 0.2 to 0.3 volts. When charging it, the typical voltages will be:

POINTS	VOLTS	
	5-HOUR RATE	20-HOUR RATE
1260	15.5	14.5
1240	15.0	14.0
1220	14.4	13.4
1200	14.0	13.0
1180	13.7	12.6
1160	13.5	12.4
1140	13.3	12.2
1120	13.0	12.0

Temperature has a considerable effect on the charging voltage. It varies about 0.21 volts for every 10^0 Fahrenheit change. Required voltage goes up as the temperature goes down. That is, a cold battery needs more voltage to charge it. Also, at lower temperatures, the energy available from a battery decreases rapidly. This drop in battery capacity is shown below:

TEMPERATURE (degrees F)	CAPACITY (%)
80	100
60	90
40	77
20	63
0	49
−20	35

Both of these effects can be the source of poor charging performance for trailer batteries mounted outside of the vehicle, where they are exposed to cold weather. A cold battery demands a heavy charging current because it is less efficient. It also needs a higher voltage to send the heavier charge current into it. On the other hand, the vehicle starting battery is warm and fully charged under the towing car's hood. Consequentially, it tells the regulator to cut down on the alternator's output. The result is a charge-starved trailer battery left out in the cold. it should be obvious, then, that you should provide some means of keeping the trailer battery warm.

Chapter 7
Generators & Alternators

Generators and alternators convert mechanical into electrical energy. The basic principle of both is the same, and the electrical energy generated in the windings is alternating in both cases. The difference is in how the alternating current (AC) is converted to DC suitable for battery charging. The generator uses a mechanical charger consisting of the brushes and commutator. The alternator uses electronic diodes.

Physically the functions of the rotating element, *rotor*, and the fixed element, *stator*, have been interchanged. In the generator the field is the fixed stator, and the generating winding is the rotor. In the automotive alternator, the field winding is on the rotor, and the generating winding is on the stator (Fig. 7-1). In some older industrial alternators the field was the stator, but this construction is no longer used.

PRINCIPLES OF OPERATION

When an electric current flows in a wire, there is a magnetic field formed in the space around the wire. This field can be increased by winding the wire into a coil, or by introducing a magnetizable iron core into the coil. The magnetic field flows thousands of times more readily in the iron than in the air. The iron conducts the magnetic field like a wire conducts electricity.

As long as the current flows, the magnetic field exists, but if the current is turned off, the magnetic field collapses and disappears. On the other hand, if a wire or coil is held in a magnetic field, there is no electricity induced in the wire. When the wire is moved in the field, however, a voltage is induced in the wire by the *changing field*, and if the wire forms a closed circuit, current will flow. This induced voltage occurs only when the magnetic field is changing relative to the wires—the wire can be moved relative to a fixed field, a field can be moved past a wire, or the field can be increased or decreased in intensity (neither wire nor field being mechanically moves). The important point is that there must be change.

In the generator, several coils of wire, wrapped around iron pole pieces cause a magnetic field in the iron when a current flows through the coils. Between the pole pieces is the rotor, also of iron so the magnetic field flows through this rotatable core, rather than through the air spaces between the poles. Another group of coils is wrapped in slots on the outside of the rotor. When the rotor is turned, these coils move through the magnetic field so an electrical voltage is induced in them. The amount of voltage induced in these coils is proportional to the strength of the magnetic field and to the rate of rotation of the rotor: double the field, double the voltage; double the RPMs, also double the voltage.

The ends of the rotating coils are carried out to BARS on the *commutator* at one end of the rotor. The electricity is picked off of the rotating commutator by *brushes*, usually made of carbon. From the brushes, the electricity is driven through the external circuit by the induced voltage of the rotating windings. The current flowing in any coil is alternating; it reverses direction or polarity every revolution of the rotor, but the commutator and brushes *rectify* the AC, and this is what flows in the external circuit.

If the commutator is replaced by two round solid rings properly connected to the rotor winding, the current flowing in the external circuit from the brushes will be alternating. This is the original design of the *alternator*. In order to charge a battery with this unit, it is necessary to convert or rectify the AC with *diodes*.

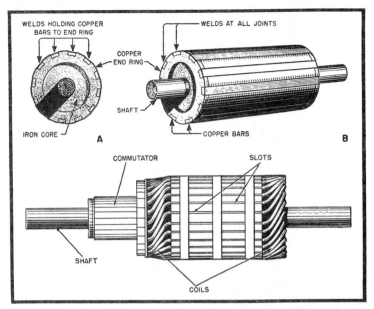

Fig. 7-1. Armatures. The top one is the type used in a DC generator. Its commutator rectified the AC voltage generated by the coils. Below is an armature that is typical of one used in an alternator. In this type the coils are located in the housing. Each copper bar is a pole of a magnet. Contrasting these two types, the DC generator has magnet field in the stator, while the alternator has its field in the rotor.

The automotive alternator has been turned inside out from this version. The field is wound on the rotor, and it is fed by two brushes and slip rings. The stator now carries a fixed, non-moving winding in which the electricity is induced by the rotating field. There are several mechanical advantages to this new construction. The brushes now only have to handle 3 to 4 amperes field current rather than the 40 to 100 or more amperes of the output. Also, it is possible to wind two more windings on the stator so voltage is induced in such a way that the outputs can be combined so as to give over a third more power from the same amount of iron and copper. The original winding is called *single-phase*; the three-coil winding is called *three-phase*.

When the alternating output from the 3-phase winding is rectified to produce DC, the output is much smoother, and if one winding or diode fails, some output is still available from the generator. In the case of a 1-phase unit, a coil or diode failure would totally disable the unit.

Also, in the interest of economy and manufacture, the rotor field can be transformed into a single bobbin coil with interleaved pole fingers on each end that is exactly equivalent to the multiple wound coils and pole pieces of the DC generator.

The generator or alternator is driven at a varying speeds by the engine since the engine speed changes to suit road conditions; a range of speeds from 800 to 8000 RPM is not unusual. Since the voltage that is generated by the machine is proportional to speed, but the battery is a constant voltage device, some means of regulating the generated voltage is needed.

The regulation is done by lowering the field current as the generator speed increases. The regulator senses the generated voltage and regulates the field current to hold it constant.

All regulators for the older DC generators and many new alternator units consist of a relay whose coil is connected across the generator output terminals. As the voltage increases, the coil attracts an iron bar, called the *armature*, and pulls it down against a spring. As the armature moves down, it opens a pair of contacts to interrupt the field current of the generator. The field current starts to fall, and with it the magnetism in the poles. The fall is not instantaneous, because of the inductance or storage effect of the field coils. As the magnetic field decreases, the generated voltage also falls. The spring then pulls the relay armature back up and recloses the contacts in the field circuit. Then the field current, the magnetism, and the generated voltage increase again. Then the relay opens the contacts again. The opening-and-closing cycle repeats often enough to hold the voltage of the generator constant at the setting of the regulator. Actually, the relay contacts do not open the field circuit fully because there is a resistor in parallel with them that always allows some current to flow.

Some regulators have a second contact that can short out the generator field to ground to better control the voltage at very high speeds.

Newer alternator regulators use transistors instead of vibrating contacts. The transistors can be connected to operate similarly to the vibrating contacts, opening and closing the circuit at a rate needed to give the proper voltage. Transistors can also be connected so as to smoothly change the current into the field as called for by the regulated voltage rather than interrupting it.

The word *generator* includes many things besides the *DC generator* commonly used in automotive service. An *alternator* is an electric generator. A battery is an electric generator. A fuel cell is an electric generator. So is an atomic pile. An alternator is a special kind of generator that produces AC. The old term for a DC generator was dynamo, but this is no longer used.

HIGH-VOLTAGE OUTPUT

Automotive alternators and DC generators are capable of putting out much higher voltages if the regulator is not used or is bypassed. However, care must be used in changing the circuits, because the higher voltages can burn out lamps, relays, instruments or other connected loads. If a high voltage is applied to a battery, more charging current can be forced into it. This is okay as long as the battery is not fully charged or not overheated by the high current.

Special care must be used in running an alternator at a high voltage to be sure that the peak inverse voltage (PIV) of the diodes is not exceeded, or they will be damaged. Diodes are available that will fit into the alternator with PIV ratings up to 600 volts. The voltage rating of the diodes furnished by the auto companies does not appear to be readily available, so raise the voltage with caution. Apparently, a number of stock alternators are capable of putting out 115 volts DC.

A DC generator will also put out considerable voltage if the regulator is bypassed. The limitation here is that the field coils may be overheated at high voltage, as this results in a correspondingly high current.

The current output rating of a generator or alternator should never be exceeded, since the current flowing in the windings causes heating. Too much current will soon burn up the unit. At these forced high outputs, the field current is also higher and contributes to the heating.

At higher voltage, the power input to the generator must be increased, so the drive belt must be exceptionally tight to prevent slippage and wear of the belt and pulleys. Too tight a belt can damage the generator bearings. If it is planned to take high power from the generator, it may pay to go to a double belt.

In the Appendix is a list of diodes, including the press-in type used in alternators with PIV ratings up to 400 volts. These diodes are available from most electronics supply companies. If you work with these diodes, be very careful not to bend the center lead near the glass seal as it can be easily cracked. Also, do not cut off the lead with a pair of pliers that closes with a snap, as the shock can break the glass also. A diode with a cracked seal may test okay, but it will soon go bad.

There are a number of devices on the market that, when attached to an alternator will allow a 110 to 120 volt DC output. These bypass the voltage regulator and apply 12 volts from the battery directly to the field of the alternator. When the alternator is turned fast enough—somewhere in the range of 4000 to 6000 RPM, it will give an output of 120 volts. Since there is normally some step-up ratio between the engine and alternator pullies, the engine will run somewhat slower.

Most alternator diodes will withstand this higher voltage when they are cool—but may break down when hot. Too much power output from the alternator could cause this heating, but practically, the output is limited by the belt drive to somewhat in the neighborhood of 10 − 14 amperes.

Above this the belt drive will slip. A loose belt will slip at a much lower speed. So, if you wish to use one of these high voltage converters, be sure the alternator belt is in good condition and well tightened. Check the belt and alternator for overheating if any continuous loading over 7 − 8 amperes is carred. Some manufacturers claim full generator output is obtainable from these devices, but as this would amount to from 10 to 12 HP input, the drive belt would not carry it without slippage and overheating.

CAUTION: the switches and thermostats of AC appliances rated at 120 volts AC may not be adequate for *frequent* operation at 120 volts DC. The interrupting capacity of these switches on DC is very much lower than for AC. Be ready at any time to pull the plug on a unit that fails to turn off when the switch is operated on DC.

Chapter 8
Wire Sizes

Practically all automotive wire is made of copper. The only three exceptions are that sometimes the heavy starter cables are made of aluminum, and some cars use a resistance primary wire between the ignition switch and the coil that is made of a special iron nickel alloy. High-voltage secondary ignition wire to the spark plugs is composed of carbon to reduce radio interference.

All wires are made in a standard series of sizes called the American wire gauge (AWG). A wire may be one solid piece or it may be made up of a bundle of smaller wires called *strands*. Solid wire is used in buildings or installations where there is no movement or vibration. Stranded wire is used where there is movement because it is more flexible and less liable to break when moved. Except for magnet wire used to wind coils and motors, wire is usually only available in the even gauge numbers.

The most important characteristic of a wire is its electrical resistance. This resistance is high for small wires and low for large ones. Wire tables give the diameter of a wire, its area, and the resistance per unit of length. The usual length is 1000 feet, but the tables given in this chapter use a more convenient length of 1 foot.

The diameter of a wire is measured in thousandths of an inch. This diameter is measured with a wire gauge or a micrometer. After some practice one learns to judge the diameter of a wire without having to measure it. Stranded wire has a somewhat larger diameter than solid wire because of the small spaces between the wires. It has a somewhat higher resistance as well.

The total resistance of a wire is proportional to the length and inversely proportional to the area. The area is given in circular mils. A circular mil is a peculiar unit of English measure similar to hogsheads, roots, or barleycorns. Use it to compare the area of two wires, but that is all. The resistance of a wire for RV applications can be found by multiplying the length of the wire by its ohms-per-foot value. The formula is as follows:

$$R = L \times r$$

where: R is the total resistance in ohms, L is the length of the wire in feet, and r is the resistance per foot from the wire table.

Remember that there must be two wires to every electrical device, so the total length of the circuit must include the two. If the device is grounded to the vehicle frame, it is safe to consider the ground resistance to be *half* of the lead resistance except where a trailer is involved. Never rely on the hitch ball for grounding a trailer. Even if there is a frame-to-frame ground wire through the electrical connector, consider the resistance of the ground circuit to be equal to the power wire resistance.

Remember, also, that the current from all of the devices in the trailer has to return through the ground wire; it will carry more current so it should be a heavier wire than the power leads.

Never use the hull of a metal boat for a ground, particularly if the boat is to be used in salt water. The stay currents from such a ground point can cause a fantastic corrosion where the current leaves the hull and enters the water.

There are two possible limits to the current capacity of a wire. One is the heating caused by the resistance. Obviously, the wire cannot be run so hot as to melt the insulation. Many insulations are cooked and hardened at a far lower temperature. The second limitation on allowable current is the voltage drop caused by the current flow in the wire. Lights, large motors, and charging circuits can be adversely affected by too large a voltage drop.

Table 8-1 shows the maximum allowable current as recommended by the American Institute of Electrical Engineers (AIEE) for marine service. The Society of Automotive Engineers has a recommendation for autos that is somewhat higher. The AIEE values are more practical for RV since the applications are more like marine service where the wires can be near flammable materials. The values in this table are based on heating, not on voltage drop. In the Appendix is another table that gives the actual voltage drop for various lengths of wire at this AIEE recommended current.

The maximum allowable voltage drop is not easy to determine. Sometimes manufacturers will recommend a given size wire for installation of their equipment. If the wire lengths given are not suitable for your installation, it is possible to figure the resistance for the size and length given, then pick a wire to maintain this value (or a lower one) for the needed length.

For want of better values, the voltage drop for charging circuits between two batteries should be less than 0.15 volts. Lights and lamps should not have more than 0.5 volts line drop. The same figure is valid for small motors, also. Large motors can tolerate 1 to 2 volts if they run intermittently; otherwise, the drop should be under 1 volt. Less is desirable, but often difficult to attain. Radio transmitters should have less than 1 volt drop if full output is to be maintained.

Table 8-1. Wire Specifications.

AWG	DIAMETER		AREA	CURRENT	RESISTANCE
	SOLID WIRE	STRANDED WIRE			
	Inches	Inches	Circ. Mils	Amps.	Ohms per Ft.
4	0.204	0.232	42,000	127	0.00026
6	0.162	0.184	26,300	96	0.00041
8	0.128	0.146	16,500	73	0.00065
10	0.102	0.116	10,400	54	0.0010
12	0.081	0.092	6,500	35	0.0017
14	0.064	0.073	4,100	20	0.0026
16	0.051	0.058	2,600	11	0.0042
18	0.040	0.046	1,600	6	0.0067
20	0.032	0.037	1,000	4	0.010

*Area to nearest hundred circular mills.

Note that the drop should include the return or ground circuit as well as the supply line. Any device that requires more than 20 amperes should have a separate ground return to the battery unless the device is within a few feet of the battery.

Also, for heavy-current devices, be sure the ground is as direct as possible to the battery. For instance, in some cars the battery ground goes to the engine, and the engine has only a light grounding strap to the car frame. Grounding a high current device to the frame will force the current through this light strap. If the strap should break or develop a high resistance, the only return for the current could be through the transmission or control rod bearings. High currents can seriously damage ball and roller bearings if it flows through them. To prevent such damage, the device return should be to the engine, not to the frame.

If the battery is grounded to the frame, then ground the device to the frame. The two most likely devices for such grounding care are electric winches and DC-AC inverters. By the way, protection of transmission bearings is a good reason to always have a good ground between the engine and frame of any vehicle.

To illustrate the use of the wire tables and other data, assume we have a trailer to be wired with the following equipment:

- Six 21 CP (candlepower) lights to be run 4 hours per day.
- One ventilating fan (4 amperes) to be run 4 hours per day.
- One water pump (6 amperes) on 1 minute, 20 times per day.
- Length of wire from the vehicle battery to these devices is 30 feet. Total run including ground is 60 feet.
- Nominal battery voltage is 13.5 volts.
- We do not want the lights to dim when the pump runs.
- Current of a 21 CP light (Chapter 10) is 1.44 amperes.

From this list of information we can make the following calculations:

LOAD	CURRENT	AMPERE-HOURS	VOLTAGE DROP
Lights	6 × 1.44 = 8.64	34.56 (4 Hours)	0.50
Fan	1 × 4 = 4	16	0.50
Pump	1 × 6 = 6	2 = 6 × 20 × 1/60	1.0

NOTES:
 1) Total load when pump is OFF: 12.64 amperes
 2) Total load when pump is ON: 18.64 amperes

To prevent flicker in the lights, the intermittent pump load should be on a separate circuit, but the fan can be on the same circuit as it is a steady load.

To find the proper wire size if the lights and fan are on the same circuit, divide the allowable voltage drop by the current drain (Ohms's law) of the lights and fan:

$$R = E/I = 0.5/12.64 = 0.039 \text{ ohms}$$

This value is for the entire run of 60 feet, so the resistance per foot is:

$$r = .039/60 = 0.00067 \text{ ohms}$$

This is about the ohms-per-foot value of AWG +8 wire listed in the wire table; so this would be the size to use. This is quite large, so see what would happen if both pump and fan motors were on the same circuit, and lights separate. The total resistance for the lights would be:

$$R = 0.5/8.64 = 0.058 \text{ ohms}$$

The ohm-per-foot of resistance of wire should be:

$$r = 0.058/60 = 0.001 \text{ ohms}$$

This is the resistance for AWG #10 wire.

In the case of the pump and fan motors, the conservation design would be to add the current of both motors to figure the wire, but with the short-on time of the pump, only the current of the fan need be considered. When the pump goes on, the fan will slow somewhat, but the effect will not be too noticeable. Therefore, we can divide the 4-ampere drain of the fan into its 0.5-volt drop to find R:

$$R = 0.5/4 = 0.013 \text{ ohms}$$

then,

$$r = 0.013/60 = 0.0021 \text{ ohms}$$

This is the approximate resistance (0.0021 ohms) of AWG #14 wire.

Now, if the water pump was used for a bait tank, or your family procedure is to take two or three consecutive showers, where the water pump would be running more or less continuously, then the voltage drop should be calculated for the current of both motors.

If a battery is installed in the trailer, and the wire runs are now only 6 to 10 feet, there would be a considerable saving in heavy-gauge wire, even for this simple installation.

The total resistance R are unchanged, but the ohms-per-foot resistance is less critical because of the shorter runs:

Lights and Fan:

$$R = 0.0396 \text{ ohms}$$

then,

$r = 0.0396/10 = .00396$ ohms

which is equal to AWG #12 wire. In the other method it was AWG #8 wire.

Lights only:

$$R = 0.0578 \text{ ohms}$$

then,

$r = 0.0578/10 = 0.00578$ ohms

which is equal to AWG #14 wire. In the other method it was AWG #10 wire.

Fan Motor:

$$R = 0.13 \text{ ohms}$$

then,

$$r = 0.13/10 = 0.0013 \text{ ohms}$$

which is equal to AWG #18 wire. In the other method it was AWG #14 wire.

BATTERY CHARGING WIRES

Let's go back to the battery charging problem of Chapter 1 to find the proper size for that 35-foot charging wire. Assuming the 0.9 volt drop is right, and 10 amperes is the charging current, the wire resistance total is:

$$R = 0.9/10 = 0.09 \text{ ohms}$$

The resistance per foot needed is:

$$r = 0.09/35 = 0.0026 \text{ ohms}$$

Referring to the wire table, the right size to use is AWG #14 wire.

Let us examine the assumption that 10 amperes is a reasonable charging rate. The ampere-hours which are used by the lights and motors totals 52.56 (34.56 + 16 + 2). To restore this charge, at 10 amperes, it would take about 6 hours to recharge the battery (more must be put in than is taken out by about 15%). If the trailer is towed each day, then this may be adequate; but if a two-night stop is made, even a 100 ampere-hour battery will be exhausted.

This much drain would call for a much larger battery or an auxiliary power plant; otherwise, the lighting load will have to be decreased quite a bit. It would be impractical to allow the car to sit with the engine running for 6 hours just to charge the battery. There is another way to handle this situation; it will be explained in Chapter 12. The idea is to use the car alternator as a high rate charger that can recharge the battery in the order of 2 hours. The circuitry is not for the novice, however.

Some increase in the charging rate during travel can be gained by going to a gauge or two heavier charging wire to the trailer, but then there is a possibility of overcharging the battery during several days of continuous running. The use of a large battery will help equalize the charge along with the longer AWG #12 wire.

If a charging wire needs to be AWG #8 or #6, it is more practical to use building wire that is readily available in these large gauges. AWG #6 can also be obtained as a welding cable. Surplus aircraft wire is also available in these large sizes.

Chapter 9
Fuses & Circuit Breakers

Fuses and circuit breakers should in no case have a rating higher than the maximum capacity of the wire in the circuit to be protected. In most cases the protector will have a much lower rating that is determined by the devices connected to the circuit.

The rating of a fuse or circuit breaker usually has to be several times the normal rating of the device because there is often an inrush (surge) current for a fraction of a second that is many times the continuous drain of the unit. Motors, incandescent lights, and transistor DC-AC inverters are examples.

No fuse or circuit breaker opens instantaneously, but most are designed to operate faster as the rated current is exceeded more or less as follows:

- 100% of the rated current can be carried continuously.
- 135% of the rated current will hold less than one hour.
- 150% of the rated current will hold 15 to 25 seconds.
- 200% of the rated current will hold less than a second.

There are special fuses called *slow-blow* types that will stand a short overload for a longer period of time; yet, they will blow about the same as normal fuses for a long time, low overload. These fuses are specially designed for motors and lights that have very high starting inrush currents.

Standard automotive fuses are not made in slow-blow versions. These fuses are SFE type with a quarter-inch diameter. The length of these fuses differ according to the ampere rating. This helps the novice from installing the improper size of fuse. A list of some of the more common sizes follows:

RATING (amperes)	LENGTH (inches)
4	5/8
6	3/4
7.5	7/8
9	7/8
14	1 1/16
20	1 1/4
30	1 7/16

Electronic fuses are made in several sizes. The commonest ones are 1/4-inch diameter by 1 1/4-inch long. They are made in both standard and slow-blow time delays. The current ratings available are from nil to 25 amperes. The lengths do not change with different ratings.

Another size of electronic fuse is 13/32-inch diameter by 1 1/2-inch long. These are rated up to 30 amperes. An aircraft fuse of the same size, type 5AG, is available up to 70 amperes, but these are sometimes hard to find. These types are the two upper-left ones in Fig. 9-1.

For other 35-, 40-, 50-, and 60-ampere fuses it is necessary to go to the industrial type FRN units that are 3/4-inch diameter and 3-inch long. One is shown in the center of Fig. 9-1, with two loose mounting clips to the right of it. Mounting clips similar to the ones in the lower-right portion are also available for these large fuses. The panel type, or in-line type mounts shown in the lower center, are available only for the 1/4-inch and 13/32-inch diameter fuses.

The two fuses shown at the lower-left corner are slow-blow types with the special heavy element needed to produce the time-delay operation.

Because of the difficulty of obtaining high-ampere fuses, circuit breakers are attractive, even though they cost quite a bit of money. At least they do not have to be replaced after every trip out, and there is no problem of having forgotten spares!

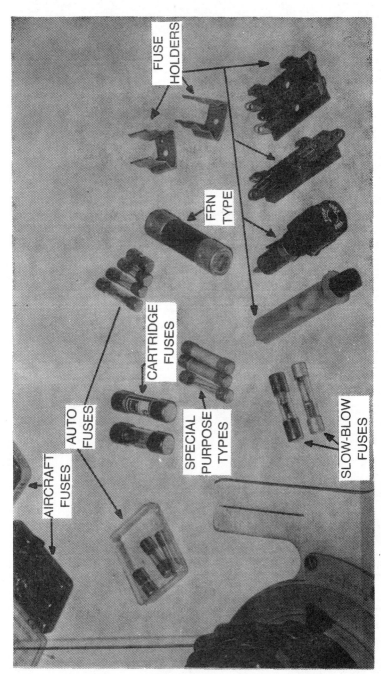

Fig. 9-1. Various fuses and fuse holder.

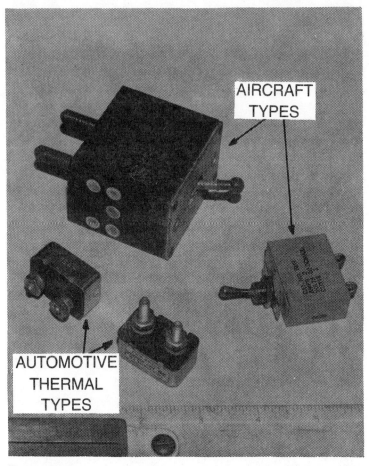

Fig. 9-2. Circuit breakers. The two aircraft-type breakers are manually resetable, while the auto-type shown have to cool down before they close.

The two breakers to the lower-left portion of Fig. 9-2 are automotive-thermal types that have to be cooled off before they reset. The others are aircraft types that can be reset by operating the handle, and they also serve as switches, if desired. The most economically priced breakers that can also be used as switches are the Wood Electric Company Models 112 and 109, available from the electronics suppliers for about $8. Current ratings are from 5 to 50 amperes DC.

Chapter 10
Electrical Loads

Unfortunately, some of the data in this chapter has to be painted with a broad brush since many of the manufacturers catalogs do not contain adequate electrical data on their products.

To be really meaningful, the data should include the no-load or idling current, the current at rated load, and the maximum surge current, if any. Also, the minimum allowable voltage at the terminals of the device should be given.

There is no piece of literature that gives even half of this data; many do not even give the rated current. Unless otherwise noted, the following data apply to 12-volt units. Six-volt units will normally draw twice the current.

EDISON SCREW BASE (HOUSEHOLD) LAMPS		
RATED WATTS	6-VOLT AMPERAGE	12-VOLT AMPERAGE
10	2.0	1.0
15	2.5	1.3
20	3.8	1.8
25	4.1	2.1

FLUORESCENT LIGHTS WITH DC — AC CONVERTOR		
RATED WATTS	6-VOLT AMPERAGE	12-VOLT AMPERAGE
15	—	1.7
30	—	3.4

FANS	
TYPE	AMPERES
4-inch	1.0 – 1.5
6 – 8-inch	4.0 – 5.0
Blower	

PUMPS		
TYPE	CAPACITY*	AMPERES
Centrifugal bait and bilge	470 GPH	4.5
w/4-foot head	750 GPH	6.0
Rubber impeller	5 GPM	7.0
w/10-foot head	7 GPM	10.0
Diaphram	4 GPM	4.5
w/8-foot head	6 – 8 GPM	6.5 – 8.5
Trailer water	–	7.5
Toilet flush (10 sec)	–	8.5
Toilet waste discharge	–	20.0

*—GPH is gallons per hour. GPM is gallons per minute.

MOTORS		
TYPE	CONDITION	AMPERES
1/12 HP	running	7.0
3/8 HP	1) no load	10.0
	2) rating	40.0
	3) starting	150.0
1/2 HP (aircraft starter with gearbox)	1) rating	60.0
	2) starting	200.0
Windshield wiper		4.0 – 6.0

RADIOS		
TYPE	CIRCUITRY	AMPERES
Auto broadcast	transistor	0.05 – 1.0
CB transceiver	transmitting (4 watts)	1.3
	receiving	0.2
Marine radiotelephone	1) transmitting (77 watts)	8.0
	receiving	0.5
	2) transmitting (101 watts)	12.0
	receiving	0.5
VHF marine transceiver	transmitting (5 watts)	1.3
	receiving	0.2
Depth finder	tube-type	3.0
	tube-type (neon)	3.0
	transistor	0.2
	3-inch chart type	0.5
Radio direction finder	tube-type	3.0
	transistor	0.2

DC – AC INVERTERS	
RATED POWER	CURRENT DRAIN
200 – 300 watts	3.0 amperes(idle) + load wattage / 10
450 – 500 watts	4.5 amperes(idle) + load wattage / 10

ELECTRIC TRAILER BRAKES	
SIZE	CURRENT DRAIN
10-inch	3 amperes per wheel

INRUSH CURRENT	
DEVICE	PERCENT OF RATED CURRENT
Incandescent lamp	500%
Small motor	300%
Medium motor	400%
Large motor	500 – 700%
Inverter	200%
Tube-type radios	200%

WINCHES		
TYPE	POUNDS OF PULL	AMPERES*
Marine anchor windlass	300	72
Converted aircraft starter	3000	100
Marine capstan	1000	60

*—current will depend on speed of rope.

ABBREVIATIONS:

*Base	MB	Miniature Bayonet
	SC	Single Contact Bayonet
	DC	Double Contact Bayonet
	DCI	Indexed Double Contact
	WG	Glass Wedge, Wire Contacts

**Shape

G	Globe
T	Tubular
P	Pear
FP	Flattened Pear

* CP   Candlepower

115-VOLT ATERNATING CURRENT LOADS	
DEVICE	WATTS
TV set	250
Refrigerator	1200
Electric iron	900
Toaster	1000
6-inch electric saw	1000
1/4-inch electric drill	250
1/4 HP motor	400
	(inrush) 1050
1/3 HP motor	450
	(inrush) 1350
1/2 HP motor	600
	(inrush) 1800
1 HP motor	1000
	(inrush) 3000
Light bulbs	

AUTOMOTIVE & MARINE LIGHT BULBS					
NUMBER	CP*	AMPERES	BASE*	SHAPE**	USE
51	1	0.22	MB	G	Indicator
55*	2	0.41	MB	G	Indicator
63	3	0.63	SC	P	
81	6	1.02	SC	P	
82	6	1.02	DC	P	Marine
87	15	1.91	SC	P	
88	15	1.91	DC	P	Marine
209	15	1.78	SC	P	
210	15	1.78	DC	P	Marine
1129	21	2.63	SC	P	
1130	21	2.63	DC	P	Marine
1133	32	3.91	SC	FP	
1154	21/3	2.63/0.75	DIC	FP	Stop/tail

12-VOLT TYPE					
NUMBER	CP*	AMPERES	BASE*	SHAPE*¹	USE
53	1	0.12	MB	G	Indicator
57	2	0.24	MB	G	Indicator
67	4	0.59	SC	G	
68	4	0.59	DC	G	Marine
89	6	0.58	SC	G	
90	6	0.58	DC	G	Marine
94	15	1.04	DC	G	Marine
158	2	0.24			
161	1	0.19	WG	T	Indicator
194	2	0.27	WG	T	Indicator
631	6	0.63	SC	G	
1003	15	0.94	SC	P	
1004	15	0.94	DC	P	Marine
1034	32/3	1.80/0.59	DCI	P	Stop/tail
1073	32	1.80	SC	P	
1141	21	1.44	SC	P	
1142	21	1.44	DC	P	Marine
1155	4	0.59	SC	G	
1156	32	2.10	SC	G	
1157	32/3	2.10/0.59	DCL	P	Stop/tail
1507	1	0.20	MB	T	Radio dial
1895	2	0.27	MB	T	Radio dial

Chapter 11
Special Boating Problems

One of the heaviest constant loads in the electrical system of a boat, and one that causes more calls for help to the Coast Guard is the drain of a bait pump. The bait tank requires a constant flow of water to keep the bait lively. The water should be changed in the order of three or four times per hour as a minimum.

In a small boat, the required bait pump capacity is in the order of 2 to 5 gallons per minute flow at a head of about 5 feet.

Head is a measure of the amount of pressure developed by the pump. It is measured as feet of water. For salt water, a 1-foot head is 0.44 PSI (pounds per square inch). For fresh water it is 0.43 PSI.

The flow of water in the pipe or hose causes a pressure loss that is also measured as head in feet of water. The pump total head is the suction lift plus the height to which the water is pumped along with the friction loss in the pipe.

The flow of 2 to 5 GPM at the usual pumping heads encountered does not represent a large amount of power, but the pumps on the market today are very inefficient, and use far too much power from the electrical system. In fact, the common rubber impeller pump seen most often, draws more power than the typical outboard engine generator can furnish. That is, the battery can never be recharged as long as the pump is running.

Since you cannot redesign your pump, it behooves you to search around for as efficient a pump as possible and not accept the first one heaved at you. The power required to lift water at 100% efficiency is given very closely by the following formula:

$$P = H \times Q/5$$

where P is the electrical power in watts required to lift water quantity Q in gallons per minute with a total head H in feet of water. For example, to lift water at a rate of 5 GPM with a total pump head of 5 feet requires:

$$P = 5 \times 5/5 = 25/5 = 5 \text{ watts}$$

Now contrast this 5 watts of power with the power requirements of a rubber impeller pump working the same load that draws 8 amperes with a 12-volt supply:

$$P = 12 \text{ volts} \times 8 \text{ amperes} = 96 \text{ watts}$$

To measure the rubber impeller pump's efficiency against the ideal, we divide its 96 watts into 5 watts. This shows an efficiency of 5.2%.

One of the more efficient pumps is a small cast-brass centrifugal pump that draws 4 amperes with 12 volts supplied. It pumps 5 GPM with a 5-foot head:

$$P = 12 \text{ volts} \times 4 \text{ amperes} = 48 \text{ watts}$$

Comparing this wattage to the ideal 5 watts shows that the cast-brass centrifugal pump is 10% efficient. This is the only pump that should be used for a single outboard engine boat if the battery is to be relied on to operate the pump and also to start the engine. Unfortunately, the water output of the centrifugal pump drops very rapidly as the pumping head increases, resulting at nil output at 7 feet.

The rubber impeller pump will keep on against a very high head. In fact, it will keep going against one so high it will burn itself up trying.

One important factor in easing the duty on the bait pump is to use large hose or pipe for the lines. The pressure loss in a pipe or hose, expressed as head in feet per foot of pipe is given in Table 11-1.

Less than a half-inch ID (inside diameter) hose should never be used, even for the shortest runs. The pressure loss

Table 11-1. Head Loss Per Hose/Pipe Size.

GPM	ID INCH SIZE			
	3/8	1/2	3/4	1
1	0.23[1]	0.07	0.02	0.01
2	1.4	0.35	0.09	0.04
4	7.0	1.9	0.32	0.10
6	23.0	4.6	0.72	0.32

1—Head loss in feet. 1 foot-head is equal to 0.43 PSI. 1 PSI is equal to 2.32 feet-head.

in a 10-foot section of half-inch ID hose at 2 GPM is 3.5 feet-head. Since the total head capacity of the small centrifugal pump discussed is 7 feet-heat, the highest actual head the pump could lift under these conditions is as follows:

Head capacity	7 feet-head
Head loss of hose	− 3.5 feet-head
Net	3.5 feet-head

And if there are turns, elbows, and valves in the line the capacity will be even less.

For rubber impeller pump, one manufacturer gives the following specifications:

PSI	HEAD-FEET	GPM	AMPERAGE
4.3	10	4.0	7
8.7	20	3.7	−
13.0	30	1.7	−

RUNNING LIGHTS

Coast Guard requirements for running lights are that they be visible for 1, 2, or 3 miles in clear weather. The variations are for different lights and the size of the boat. The size of light bulb needed to give this visibility depends on whether the fixture has a plain or Fresnel lens. Most marine

Table 11-2. Visibility of Fresnel Lens Fixtures.

COLOR	CP*	MILES VISIBLE	6-VOLT NO.	12-VOLT NO.
Red	6	1	81	90
Green	15	1	87	1004/94
White	4	2	81	68
White	6	3	87	90

Note: Light bulbs given in this listing meet or surpass requirements.
*—candlepower.

Table 11-3. Visibility of Plain Lens Fixtures.

COLOR	CP*	MILES VISIBLE	6-VOLT NO.	12-VOLT NO.
Red	21	1	1129	1142
Green	21	1	1129	1142
White	15	2	87	1004/90
White	21	3	1129	1142

Note: Light bulbs given in this listing meet or surpass requirements.
*—candlepower.

bulbs are double contact because a separate ground wire is needed in a wood or glass structure. This type is also recommended for a metal hull to minimize corrosion. Tables 11-2 and 11-3 give the 6-volt and 12-volt bulbs for the required visibility. Some bulbs may not be exact equivalents, as they may not be a bulb of the necessary brightness.

In the event that electrical light fixtures are single-contact types, the double-contact equivalents are as follows:

DC	SC
81	82
87	88
90	89
1004	1003
1142	1141
68	67

WIRE TERMINALS

Crimp-type wire terminals should be avoided where they are exposed to salt water. The salt gets into the crevices of the terminals and causes a special kind of virulent corrosion. It is better to use solder, even in addition to crimping, or dip the terminal in an asphalt-type paint to seal the crevices.

Far worse is to use a copper crimp terminal on stranded aluminum wire. Such a connection can develop an open circuit with no visible appearance of damage.

Water-emulsified asphalt roofing paint, a material that looks like heavy brown chocolate, is an excellent material for painting electrical connections. It can be applied to damp parts, and when it dries it forms a very waterproof seal. It is, however, readily soluable in gasoline or diesel oil, and it will bleed through oil based paints.

It is an excellent protection for battery terminals and lugs. But it should not be allowed to dry before the lug is assembled to the battery post as the dry film is a good insulator. Assemble the lug to the post with the paint wet, then it will be squeezed out of the contact area and will fill all the voids in the joint.

Coat the bases of exposed light bulbs with waterproof grease before inserting them in the sockets, and there will be less corrosion and problems of removing a burned-out bulb. This comment also applies to boat trailer bulbs and plug-in connector.

Chapter 12
Electric Trailer Brakes

One of the better kept secrets of trailer life seems to be the electrical characteristics of trailer brakes. There have been several articles on the mechanics of their operation, and one trailer handbook gave a description of how to tune them electrically with the step resistor normally furnished by the manufacturer.

Kelsey Hays brakes consist of an electromagnet suspended on an arm that operates the normal brake shoes with an eccentric cam. The cam expands the brake shoes in the same way a hydraulic piston operates ordinary car brakes.

As electrical power is applied to the solenoid, it drags against a steel disk inside of the drum. The drag is proportional to the current flowing, so the force exerted by the lever on the brake shoes is also proportional to the electrical current.

The current is controlled by a variable resistor in the controller of the towing vehicle. This resistor is about 8 ohms when the contact is first made in the controller, and it decreases to zero as the brake pressure increases to a maximum.

The resistance of one 1.5 × 10-inch brake solenoid is about 3 ohms. So in a 12-volt system, with the controller set for full brake effort, the current per wheel is 4 amperes

(Ohm's law). When the controller resistance is first engaged, the current for a two-wheel system is about 1.2 amperes; and for a four-wheel system, about 1.5 amperes.

The maximum brake current has been designed to just lock the brakes at a certain trailer weight per wheel. For any lesser weight, the wheel will skid. This means that less current is needed per wheel for a lighter trailer to just reach the skidding point. There is an adjustment in the brake controller to change the rate of buildup of the current, but the final current is always the same, and it is too much for a light trailer.

To correct this problem, a series resistor is added to the brake circuit to limit the maximum current. This device, normally furnished by the brake manufacturer, is a board with three fixed resistors and a chart showing how to connect the resistor for different weights of trailers. Since the only available changes are in steps, it is possible to have a combination of trailer weight and braking effects that are not satisfied by any resistance step.

There is a simple way of obtaining an infinite variation in this adjustment: Use a high-wattage variable resistor or rheostat in the brake circuit. There are two styles of these resistors available, and they are made by several manufacturers. One is a rotary resistor (rheostat) that can be adjusted with a knob. The other is a tubular fixed resistor with a sliding tap that can be moved along the body and locked with a screw. Figure 12-1 shows both types of resistors as made by the Ohmite Corporation.

With two-wheel trailers, for either large or small brakes, a 4-ohm resistor of 100 watts power rating will give a full range of control. On four-wheel trailers, either brake size, use a 2-ohm resistor.

If you have an older trailer with 6-volt brakes and you intend to use it on a 12-volt vehicle, use two 2-ohm, 75-watt resistors in series, or a 175-watt, 4-ohm unit. If you want a rheostat control, use a 2-ohm, 100-watt slider resistor in series with a 100-watt, 2-ohm rheostat.

The current in 6-volt brakes is about twice that in 12-volt systems. So, with four wheels, it may be necessary to parallel the resistors rather than connecting them in series. Connect one end of each resistor to one another. Then connect the

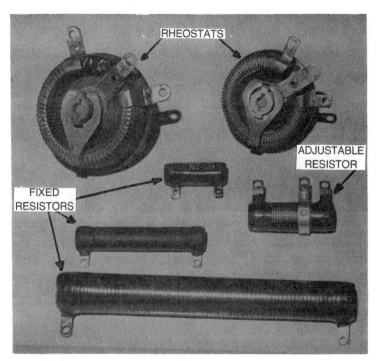

Fig. 12-1. Rheostats and power-handling resistors.

slider terminals of these resistors together. This places them in parallel. Leave the other end of the resistors unconnected. Connect the rotor tap of a rheostat to the slider terminal of the slider-resistor pair.

Either style of resistor is satisfactory. If the trailer load is not changed often or very much, the slider type is the most economical. If the trailer load changes or braking conditions vary, the rotary type may be preferable.

There is some heat generated in the resistor, so it should be mounted away from any heat-sensitive material. The resistor must be connected between the controller and the trailer brakes. Do not connect it between the battery and the controller because this line carries stop-light current as well as brake current. A resistor in this line would cause improper operation of both the brakes and stoplights.

The resistor body is made of ceramic, so it should be mounted in a place where it is protected from damage. Also, the terminals should be protected from damage. In addition, they should be protected from accidental shorting by metal objects like tools.

To set the series resistor, put the control adjustment about at midrange with the series resistance at zero. Try the trailer brakes. If they grab, the wheels skid, or the tow bucks badly, add series resistance until a smooth but positive stop is obtained. There will always be some noise or jerk in the hitch when the trailer brakes are applied, since when this is done the load on the hitch is reversed. Any slack causes some noise.

To be sure the trailer brakes are operating, apply them with the hand lever on the controller without touching the vehicle brakes. There should be a definite braking effect. Whether the braking needs to be strong enough to skid the trailer wheels depends on your driving habits and the road condition.

Never allow the trailer brakes to skid excessively because directional control will be lost, the trailer may fishtail, or on slippery roads even jackknife. An ideal setting is one that lets the trailer brakes carry a little more of their share of the stopping load so that the rig is kept in alignment. Too low of an adjustment of the trailer brakes can also cause fishtailing or a jackknife.

Be very careful of all the wires and connections in the trailer brake system. A failure here can cause an accident. Protect the wire where it goes through holes in any metal with grommets or tape. Tie the wires down where they are exposed to wind to prevent whipping and chaffing against a sharp edge. Fasten the wire to the backside of the trailer axle so no stones can cut it. Do not allow any loops of wire to hang below the axle where they can snag. Crimp connectors should be crimped *and soldered*. The fine holes and voids in a crimp connection are an invitation to a bad form of corrosion called crevice corrosion. Once some salts and moisture have gotten into such a place, the corrosion continues as long as there is the least amount of moisture present. So, solder the connections and cover them with a tight layer of tape.

If you rewire brakes, use AWG #16 wire to each wheel and at least AWH #12 wire for the run up to the controller. Be sure to wire all wheels in parallel. Have a good ground connection between the trailer and vehicle frames.

There is some question as to whether a fuse should be used in a trailer brake circuit. There is too much danger of a fire from shorting the long wire—use a fuse—20 ampere for 12-volt brakes and 30 ampere for 6-volt brakes used on a 12 volt system. Carry several spares.

Several sizes of these resistors are shown in Fig. 12-1. The upper-left one is a 75-watt rheostat, while the upper-right one is a 50-watt rheostat. In the center is a 12-watt fixed resistor, and on the right-center spot is a 25-watt slider-adjustable type. the lowest one is a 100-watt fixed resistor with a 50-watt size above it.

A convenient place to mount a resistor is inside of the trailer frame, as shown in Fig. 12-2. In this case, a 50-watt

Fig. 12-2. Convenient location of a series resistor for trailer electric brakes. In this case, a 50-watt rheostat and a 50-watt slider resistor are connected in series to obtain the needed 75-watt power capability.

109

rheostat was the largest that would fit inside of the channel. Since a 75-watt resistor was needed, a 50-watt slider-adjustable resistor was used in series to obtain the total wattage. The rheostat is 1 ohm, and the slider resistor 2 ohms. The available adjustment was then from 0.25 to 3.00 ohms total. Slider-adjustable resistors seldom will go the entire ohm range of adjustment since the sliders cannot be positioned all the way up to the terminals. Rheostats will adjust from zero to the maximum rated value of the resistance.

Table 12-1. Fixed and Slider Power Resistors.

OHMS	MAXIMUM AMPERES	OHMS	MAXIMUM AMPERES
25-WATT 2" Long by 9/16" Diameter			
1	5.00	5	2.25
2	3.54	10	1.58
3	2.88	15	1.29
50-WATT 4" Long by 9/16" Diameter			
1	7.07	5	3.16
2	5.00	10	2.23
3	4.07	25	1.41
4	3.53		
75-WATT 6" Long by 9/16" Diameter			
1	8.66	5	3.87
2	6.12	10	2.74
3	5.00	15	2.24
4	4.33		
100-WATT 6 1/2" Long by 3/4" Diameter			
1	10.0	5	4.47
2	7.07	10	3.16
3	5.77	25	2.00
4	5.00		
175-WATT 8 1/2" Long by 1 1/8" Diameter			
1	13.2	5	5.92
2	9.35	10	4.18
3	7.63	25	2.64
4	6.60		

Table 12-2. Power Rheostats.

OHMS	MAXIMUM AMPERES	OHMS	MAXIMUM AMPERES
\multicolumn{4}{c}{25-WATTS — 1 9/16" Diameter by 1 3/8" Deep}			
1	5.0	6	2.0
2	3.5	8	1.77
3	2.9		
\multicolumn{4}{c}{50-WATT — 2 5/16" Diameter by 1 3/8" Deep}			
1	10.0	6	2.9
2	5.0	8	2.5
4	3.5		
\multicolumn{4}{c}{75-WATT — 2 3/4" Diameter by 1 3/4" Deep}			
0.5	12.5	3	5.0
1.	8.7	5	3.9
2.	6.1	10	2.7
\multicolumn{4}{c}{100-WATT — 3 1/8" Diameter by 1 3/4" Deep}			
0.5	14.0	3	5.6
1.	10.0	5	4.5
2.	7.1	10	3.2

Tables 12-1 and 12-2 show various sizes of fixed, slider, and rheostat power resistors. Each table is grouped into wattages. When selecting the proper size of resistor for these trailer brake applications, allow yourself about 10 – 20# for unforseen loads, etc. It is better to use a resistor that is underpowered than one that is overpowered.

Chapter 13
High-Current Charging

A vehicle alternator is capable of charging a battery at a fairly high rate if the influence of the regulator is removed or modified. Also an external battery can be charged, independently of the vehicle battery and ignition system by a relatively simple switching circuit. An external battery is defined as any other battery of the same voltage as the vehicle system, that is not connected in any way to the vehicle except through the special charging circuits to be described.

The simplest system, and the one least liable to cause damage to the alternator or regulator is a simple modification of the Class 2A charging system. If a double-pole, double-throw (DPDT) switch is added to the circuit as shown in Fig. 13-1, the regulator sensing point and the generator output can be transferred to an external battery. By bringing the voltage-sensing lead out separately and connecting it directly to the battery terminals, the length and size of the charging line is not important. It can be sized only with regard to adequate current carrying capacity and the voltage drop need not be considered.

This connection is safe, because the regulator still controls the battery charging, and it will taper off as the battery voltage builds up. It may, however, require a relatively long time to charge the battery fully. An ammeter in the line is a useful accessory, since the charging can be discontinued when

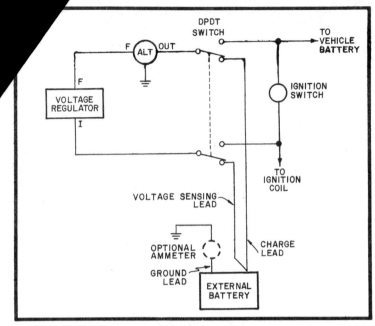

Fig. 13-1. Class 2A system modified for high-current charging.

the rate goes below 3 or 4 amperes. It is not economic to run the engine to charge at this low value.

When a Class 2A system uses an indicator light, this light must be disconnected for external charging. The proper circuit for a Delco system is shown in Fig. 13-2. Note that the indicator light terminal may be identified as I or 4. Ford systems use the letter I. Do not open the S terminal line to the Ford generator; it operates the field relay.

Class 1 and Class 3 systems, whether the older 6- and 12-volt generator types or the newer Delcotron integral alternator-regulator units, can charge an external battery using the circuit of Fig. 13-3. Only a single-pole, double-throw (SPDT) switch is needed, except for the newest Delcotron with three terminals: BAT, 1, and 2. In this instance, another switch point is needed to transfer the voltage-sensing line 2 to the external battery. Point 1, beyond the resistor in this line, must also be switched. The resistor normally connects to the ignition switch, but since there are other internal connections in the regulator between this line and the BAT terminal, there is a possibility of trouble if the line is not transferred.

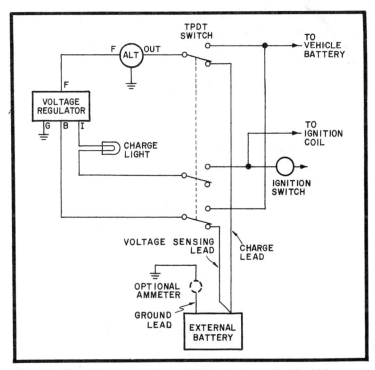

Fig. 13-2. Class 2A system with an indicator light modified for high-current charging.

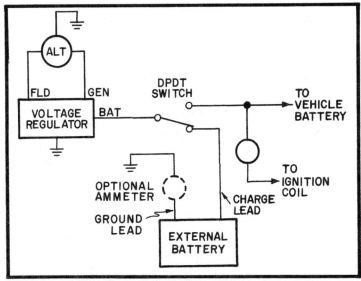

Fig. 13-3. Class 1 and 3 systems modified for high-current charging.

This line may be difficult to locate, and it is not recommended that these 3-terminal units be used on an external battery unless you know exactly what you are doing in this circuit.

Be careful with this circuit connection on older DC systems. Watch the polarity of the battery, since some DC systems have a positive ground. Also, if the charging system has an indicating light, another pole on the switch must be added to disconnect the light.

The switches for all of these circuits must be able to handle at least 15 amperes. Suitable switches are:

- DPDT
  1) Cutler Hammer: 8690K1, 7560K6, 7560K5
  2) Arrow Hart: 82610
- TPDT
  1) Cutler Hammer: 7602K2, 7612K2
  2) Arrow Hart: 82616

Wire size can be AWG #12 to #14 for the charging line to the battery, with the voltage-sensing lines AWG #18 or #20.

Class 1, 2, and 3 charging systems can be still further modified to give higher, sustained charging rates. But the modification requires that the charging be *closely supervised* because the battery is not protected from overcharging and overheating.

Also there are considerable detail differences in how the generators or alternators can be connected. Since high voltages can be generated by this system, any transistors, lights, and relays must be isolated to prevent damage.

The basic control is obtained by replacing the voltage regulator with a manually adjustable field rheostat in the generator field circuit. The maximum current of the generator should not be exceeded. In fact, it is wise not to go over three-quarters of the rated current at any time, since there is extra heat developed in the field circuit with this method of control. The vehicle will not be moving, so there will be less cooling of the generator.

Voltage of the generator should never be allowed to build up excessively either. Always turn the rheostat to the maximum resistance (minimum field current) before shutting down or starting up. The battery should be watched for heating and heavy gassing and the charging rate cut down when these occur. An ammeter in the charging line is a must for this method of control.

In the older Class 1 systems, a diode is needed in the charging line as shown in Fig. 13-4. This is because it is possible to cut the generator voltage below that of the battery, so that reverse current will flow into the generator, attempting to run it as a motor, overheating it or possibly even reversing the polarity. Alternators do not need this diode as they already have internal diodes.

Do not attempt to use the self-contained Delcotron units of any type for this style of fast charging.

Caution: In all of these systems, the transfer switch should not be thrown to the external battery when the engine is not running since the battery will discharge through the alternator field. Also, shift the switch to the vehicle battery before stopping the engine. Do not disconnect the external battery before the switch is back to the vehicle battery, since a switching surge could break down the alternator diodes.

Note that in all of these systems, the engine ignition current must be furnished by the vehicle battery. This is normally no problem for the length of time the external circuit is being used.

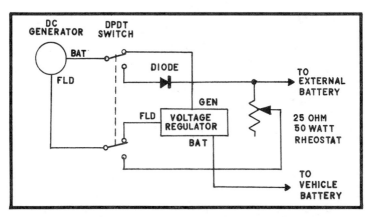

Fig. 13-4. Class 1 system that has its voltage regulator bypassed with a rheostat for even higher charging current. Note the polarity of the diode. This is for a negative-ground system. When a positive-ground system is encountered, reverse the diode.

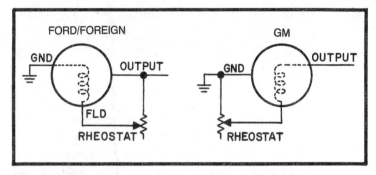

Fig. 13-5. Two methods of connecting the field windings of DC generators in Class 1 systems. The left-hand diagram applies to Ford and some foreign cars, while the right-hand one applies to GM.

Since there may be some problem of obtaining switches heavy enough to break the high-charging current possible with these systems, never switch at the maximum current. Then a heavy-duty switch rated at 25 amperes or so will serve.

To modify a Class 1 charging system for high output, follow the diagram of Fig. 13-4. Note that one basically substitutes a rheostat for the regulator, and transfers the output to the external battery. The diode polarity is illustrated for a negative-ground system; if the positive is grounded, the diode must be reversed. If there is an indicator light connected to the regulator, this connection must be opened with an additional switch contact.

In addition to the polarity differences, you will find that in the older DC systems, there are two ways of hooking up the generator field, as is shown in Fig. 13-5. In one method, common to Ford and some foreign cars, the field has one side internally grounded and the other one brought out to a terminal that must be connected to the output of the generator through the rheostat to work properly. In the other method, common to General Motors, the inner terminal of the field is connected to the output of the generator, and the field terminal of the generator has to be connected through the rheostat to ground.

If you are not sure which connection is used on your generator, try one. If the generator builds up voltage as rheostat resistance is cut out (reduced), the connection is proper. If nothing happens, try the other. No harm is done by the wrong try, but be careful not to connect the rheostat between the generator output terminal and ground. The output terminal is always a larger diameter screw than the field terminal. The field terminal, if marked will be *F* or *FLD*.

Of the Class 2B systems, the Ford is the easier to modify as there are only two versions: one with an indicator light, and one without. The two necessary circuits are shown in Fig. 13-6. Normally the one without the light does not have an *S* terminal, but if there should be one on your unit, it must be switched open for the external circuit to be used.

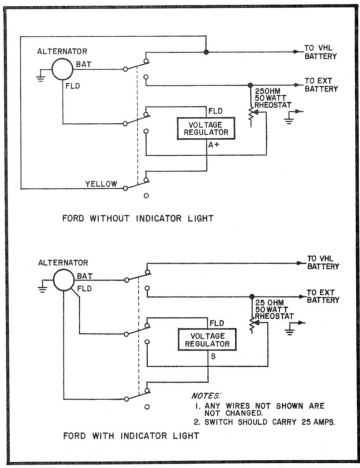

Fig. 13-6. Superhigh-current charging circuits for Fords in the Class 2B systems. The top diagram applies to Fords without an indicator light, while the one below applies to those with one.

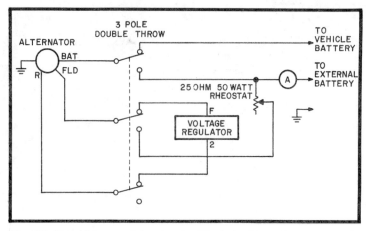

Fig. 13-7. Superhigh-current charging circuit for Delco Class 2B systems with the mechanical voltage regulator and field relay in one box. This diagram also applies to two other versions of this same type: one with a two-unit regulator containing a transistor and field relay and another with a 3-terminal transistorized regulator.

Delco has numerous variations of their charging circuits, and only the more common are shown here. However, hints are given for the less common ones.

The older, mechanical regulator was called a *two-unit* type, having a voltage regulator and field relay in *one* box. The terminals were F, 2, 3, and 4. Terminal 2 went to R on the generator, and must be switched out as shown in Fig. 13-7.

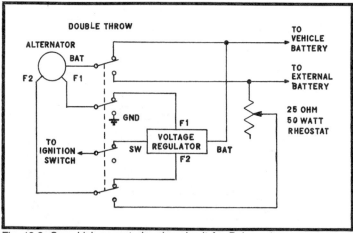

Fig. 13-8. Superhigh-current charging circuit for Delco using a mechanical-transistor voltage regulator with an internal field relay.

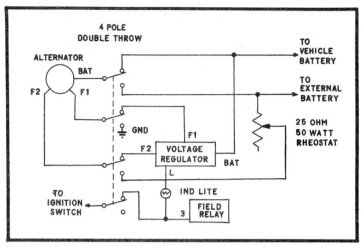

Fig. 13-9. Superhigh-current charging circuit for Delco using a mechanical-transistor voltage regulator with an external field relay.

Another version of this used a transistor as the regulator and a mechanical relay for the field with both devices in the same box. The terminal marks were the same as the one-box, two-unit type. This version can be modified in the same manner using Fig. 13-7. The indicator light line does not have to be opened for these two units.

A 3-terminal transistorized regulator, with terminals marked F, 3, and 4 exists that can be connected as in Fig. 13-7. Substitute terminal 3 for 2 of the diagram, and opening terminal 4 if there is any connection to it.

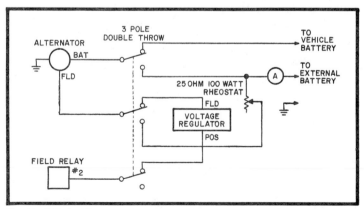

Fig. 13-10. Superhigh-current charging circuit for heavy-duty Delco systems with a separate field relay.

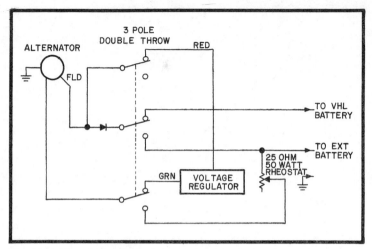

Fig. 13-11. Superhigh-current charging circuit for Motorola systems.

Delco also produced some systems where the alternator field is not grounded. There are two field terminals, *F1*, and *F2*. One version has a mechanical field and voltage relay along with a transistor in one box; the other version places the field relay in a separate box. The hookups for these two units are shown in Figs. 13-8 and 13-9.

The Delco heavy-duty regulator is a transistorized unit with a plug receptacle instead of terminals. The pins are labeled *NEG, POS,* and *F*. The field relay and light relay are separate boxes. The necessary modification of this system is shown in Fig. 13-10.

The Motorola connections are shown in Fig. 13-11. There are no terminal or plug designations, you must go by the color of the wires for this unit.

Chapter 14
Auxiliary Power Units

Auxiliary power units (APU), driven by a small engine are becoming deservedly popular. They make the RV quite independent of battery limitations and AC supplies.

There are six basic types of these plants. Based on the electrical output characteristics, they are as follows:
- Type 1—60 Hz (cycles per second) AC, 120 or 240 volts, single-phase; or 200/108 volts, three-phase. No DC outputs.
- Type 2—60 Hz AC, 120 or 240 volts, single-phase, with 6- or 12-volt DC output.
- Type 3—6- or 12-volt DC output only.
- Type 4—High frequency AC, 120 volts, single-phase in the range of 180 to 400 Hz. No DC outputs.
- Type 5—High frequency AC, 120 volts, single-phase, and 12-volts DC.
- Type 6—24 to 120 volts DC only. Usually military surplus units or older 32-volt farm units.

The units that have a 6- or 12-volt DC output can be connected to a vehicle charging circuit of corresponding voltage in a very simple manner: a single diode in the *hot* line will allow the unit to be started at any time and brought up to speed to take over any or all of the load. The circuit is shown in Fig. 14-1. The amount of load taken by the unit may normally have to be adjusted by changing the engine speed. However, this can be a problem if it is a combination AC/DC unit since speed changes will also affect the AC voltage and frequency. If the AC is not to be used while the DC is on the line, there will be no conflict. One type of load at a time is probably the only practical way to use the APU, since there is usually no way of independently controlling the AC and DC outputs.

If the APU happens to have more than adequate DC output while running at the proper speed for the correct AC output, it is possible to add a series variable resistor to limit the battery current. Such a resistor would have to be quite large as it will have to carry a high current.

There are large rheostats, similar to the ones illustrated in Fig. 12-1, that are available; or, a resistor might be made up of stainless-steel stripping or similar material. Regular iron strapping does not have enough resistance to be used. Stainless strapping—0.5 × .012-inch—has a resistance of almost exactly 0.01 ohms per foot.

Most AC/DC units derive the DC power from rectifiers inside of the unit. Some units use part of this power to excite the field winding. Other small units use permanent magnets and do not require any field current. All older DC generators used a field that requires power. If any of these units that require an internal field excitation current is connected across a battery with the engine stopped, current will flow and the battery will be run down if the unit is not disconnected. Also, the field may be burned up because there is not very much in the generator when it is not running. This is the reason for inserting the diode in the charging line; it blocks the field current flow. Some permanent-magnet generators do not need this diode. But, no harm is done if it is used, and it is a good safety precaution.

All units should incorporate an ammeter in the load circuit so the output current can be monitored.

Suitable diodes for this service are like the larger ones shown in Fig. 14-2. They should be rated at least 50 volts PIV at least 25 amperes forward current. If a charging load over 20 amperes is contemplated, use a diode rated at least 25% higher in current. Any diode should be mounted on a heat-dissipating plate. Up to 25-ampere applications can be a 3/16-

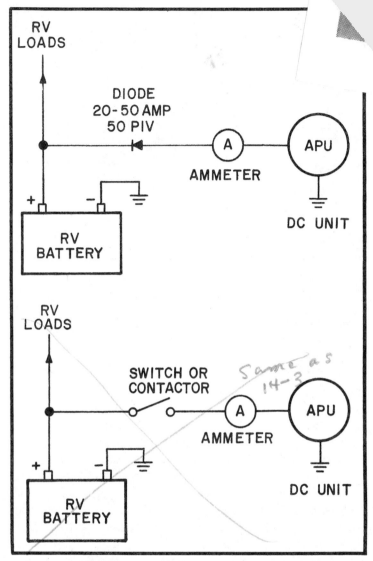

Fig. 14-1. Method for hooking up a DC APU to the RV using a diode.

or 1/4-inch thick aluminum plate, 4 to 6 inches square. When service is over 25 amperes, use one of the larger commercial heat dissipators, such as the Thermolloy 6400 B, or similar, obtainable from any of the electronics supply houses.

Since APUs usually do not have voltage regulators, the battery should be monitored in accordance with Chapter 6

Fig. 14-2. Typical diodes. Stud-mounted diodes are designed to handle more current by dissipating more heat.

while it is charging. A single-pole switch or contactor can be used to connect the APU to the RV system, as is shown in Fig. 14-3. The switch or contactor should only be closed after the APU is running and the generated voltage is equal to or slightly greater than the battery voltage. Remember, the switch should be opened before the APU engine is stopped.

If one fails to open the switch, a DC generator acts as a motor to drive the engine. Its polarity can be reversed in the process, so the next time the APU is started, the generator will fail to build up voltage, or it will build up in the wrong direction. An alternator field will draw current, but polarity will not reverse. Diodes in the line will prevent any problems if the engine stops, and they are definitely the safest way to connect the APU to the RV system.

If the APU will not supply power into the RV system or battery, check the voltage between the power unit terminals and compare it with the battery voltage, being careful to check polarity as well as value. If the APU voltage polarity is proper, but less than the battery voltage, the engine may not be running fast enough. If there is a large difference in voltage, suspect a reversal of the generator polarity, a defective diode within the APU, or other internal generator troubles.

Type 1 and 4 units can be used to feed the battery circuit by using a household battery charger of the proper voltage between the AC output of the APU and the RV's DC system. The current should be limited to the rating of the charger. The battery charger already incorporates a diode or similar rectifier, so no additional diode in series is required. Connect the charger directly to the RV's DC system—being careful to observe proper polarity—positive lead of the charger to the positive terminal of the battery. Figure 14-4 shows the circuit to be followed. It is probably not worthwhile to use a charger of less than 6 amperes capacity in this way: a 10–15-ampere unit will charge the battery much faster and will share more of the load during use. There are some new units that have been developed to carry most of the RV load when plugged into an AC power source. They will work with an APU, but the input voltage should be checked to see if it is close to the rating of the AC-DC converter, as they are called.

Type 6, high-voltage DC power units, cannot be connected to the DC vehicle system without rewinding or other modification, so they are not recommended.

The AC output of the APUs can be connected to the AC electrical system by using a double-pole, double-throw transfer switch as shown in Fig. 14-5. The cabin load is either on the APU or on the outside power system, but not on both.

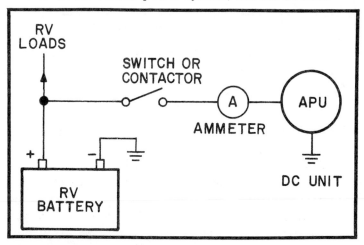

Fig. 14-3. Method for connecting a DC APU to the RV battery through the use of a switch or contactor.

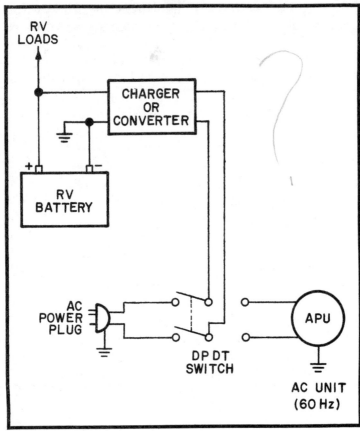

Fig. 14-4. Method of connecting an AC APU to the RV's DC systems. The APU powers a battery charger or converter to charge the battery. Or when convenient, the charger can be plugged into a household AC outlet.

Warning: Under no circumstances try to parallel an AC APC either with another one or with the AC power line.

It is impractical to try to parallel APCs to obtain more output. Refer to the comments concerning dual-engine systems.

When an attempt is made to parallel units with AC outputs (as contrasted to automotive alternators which actually have a DC output because of the internal rectifying diodes) and limited-power engines, the generator will not remain synchronized unless a special winding is incorporated in the rotor. They will hunt and surge until something gives: a stalled engine, blown fuse, or other damage.

DC generators cannot be paralleled either, unless blocking diodes are used in series with the outputs. Otherwise, if one engine slows down or the voltage drops in its generator for some reason, the generator will attempt to run as a motor. Special requirements are demanded of voltage regulators for paralleling that are not incorporated in automotive units.

Switches recommended for connecting the APU to the RV electrical system are the Cutler Hammer 7562K23 or Arrow Hart 82608, or similar totally enclosed, double-pole, double-throw, open-center switches. Do not attempt to use an open knife switch or the household three-way or four-way switches. An open switch is a shock and short-circuit hazard. The three-and four-way switches do not have the proper switching sequence for this service.

If your APU is accessible only from outside of the cabin, it may be worthwhile to add a remote stop button. For the type that already has a remote stop button, any number of other buttons can be added in parallel. A common doorbell button is satisfactory.

Run two wires, one from the engine frame and one from the contact on the stop button that goes to the breaker points of the magneto. Sometimes, there are already two wires to

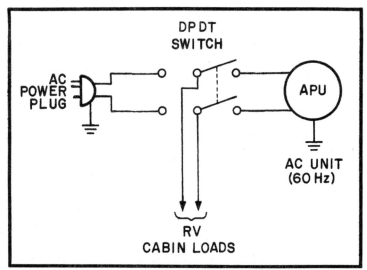

Fig. 14-5. Method for connecting an AC APU to an RV that has a AC voltage system.

the button. Simply parallel your new button across these wires. The switch grounds the magneto or breaker points to the engine frame to kill the ignition spark. The voltage in this circuit is high, so be careful to use good wire that will not get leaky when it is wet, make good contacts, and use at least AWG #18 wire.

If the engine uses a spark plug shorting spring to stop, it will be necessary to go into the magneto or the points breaker box to attach a wire to the points. Attach it to the same connection where the wire or insulated tab of the capacitor goes. This point is insulated from the frame, and there will be a wire from this point to the ignition coil. In fact, if the engine has an accessible coil, attach your stop button wire to the point side of the coil. In this way the button shorts out the points when it is closed. Be careful not to attach it to the battery or ignition switch side of the coil or you will short the battery and burn up your little project, if not more.

Appendices

Appendix A—Voltage Drop in Wires at Rated Current

Appendix B—Electrical Instruments Listing

Appendix C—Diode Listings

Appendix D—Relay Listings

Appendix E—Directory of Manufacturers

Appendix F—Mail-Order Houses

Appendix A
Voltage Drop in Wires at Rated Current

The wire voltage drop is at rated maximum current for run shown.

AWG	AMPERAGE	VOLTAGE				
		5 ft	10 ft	15 ft	20 ft	25 ft
4	127	0.16	0.33	0.50	0.66	0.82
6	96	0.18	0.39	0.59	0.78	0.99
8	73	0.24	0.47	0.70	0.94	1.18
10	54	0.27	0.54	0.81	1.08	1.35
12	35	0.29	0.58	0.87	1.06	1.32
14	20	0.21	0.42	0.63	0.84	1.05
16	11	0.21	0.42	0.62	0.83	1.03
18	6	0.21	0.41	0.61	0.82	1.02
20	4	0.20	0.41	0.61	0.81	1.01

Appendix B
Electrical Instruments Listing

VOLT-OHM-MILLIAMMETERS

Manufacturer   Model No.

RCA   WV-516A
Eico   4A3
Micronta   22-4027

VOLTMETERS

Emico   RF 2 1/4 C 20 volt
Shurite   850
Hoyt

AMMETERS

Emico   RF 2/14 C 10 ampere
Surite   850 25 ampere
Hoyt   20-0-20 ampere

HOLD-ON AMMETERS

Emico   750D5 75-0-75 ampere
J. C. Whitney   75-1984 60-0-60 ampere
Snap-on Tools

NOTE: Meters available from electronic supply companies and many automotive supply stores.

Appendix C
Diode Listings

5-AMPERE DIODES

50 Volts PIV	100 Volts PIV
1N1612	1N1059
1N2228	1N1065
	1N1071
	1N1089

12-AMPERE DIODES

1N1199	1N1200
1N2576	1N2577
1N2587	1N2588
	1N2599

25-AMPERE DIODES

1N3491	1N1454
1N2154	1N3492
	1N1192

50-AMPERE DIODES

50 Volts PIV	100 Volts PIV
1N411B	1N2447
1N2172	1N1462
1N2458	1N412B
1N2128	1N2173

PRESS-IN DIODES FOR ALTERNATORS

1N3491/Motorola MR322: 50 Volts PIV, 25-ampere
1N3492/Motorola MR323: 100 volts PIV, 24-ampere
1N3495/Motorola MR326: 300 volts PIV, 25-ampere
Motorola MR327: 400 volts PIV, 25-ampere

HOT-CARRIER DIODES, PRESS IN CASE

Half of the forward voltage drop of ordinary diodes:
Motorola MBD 5500: 20 volts, PIV, 50-ampere

Appendix D
Relay Listings

10-Ampere, 12-Volt Contacts—12-Volt Coils

Potter Brumfield	SPDT	5D15
	DPDT	11D15
Guardian	DPDT	122S-2c-12D

20-Ampere, 12-Volt Contacts—12-Volt Coil

Guardian	SPDT	PR-5AY
	DPDT	PR-11AY
RMB	SPST	60-902

50-Ampere, 12-Volt Contacts—12-Volt Coil

Magnecraft	SPST	W88KDX-2

100-Ampere, 12-Volt Contacts—12-Volt Coils

RMB	SPST	70-902
	SPDT	70-910

NOTES:
1) RMB available from Neward Electronics, others from any electronic mail-order company.
2) Ohmite high-power resistors and rheostats available from any electronic mail-order company.

Appendix E
Directory of Manufacturers

Most of these manufacturers will not sell one or two units by mail, but if you are unable to locate a dealer in your area, they will supply names of sources. Many of the products are handled by the mail-order companies listed elsewhere in this Appendix.

MANUFACTURER	EQUIPMENT
Bussman Mfg. Div. University and Jefferson St. Louis, Missouri	Fuses
Eico 283 Malta St. Brooklyn, New York	VOMs
EMICO 8th and Chestnut Streets Perkasie, Pennsylvania	Meters
Guardian Electric 155 W. Carrol Ave. Chicago, Illinois	Relays
Hoyt Electrical Instruments 45 – 47 Washington Penacook, New Hampshire	Meters

MANUFACTURER	EQUIPMENT
Kelsey Hays Brakes Romulus, Michigan	Brakes
Magnecraft Electric 5575 N. Lynch St. Chicago, Illinois	Relays
Motorola Semiconductors 5005 E. McDowell Phoenix, Arizona	Diodes
Ohmite Mfg. Co. 3601 Howard St. Skokie, Illinois	Resistors
Potter Brumfield 1200 E. Broadway Princeton, Indiana	Relays
RCA 415 So. Fifth St. Harrison, New Jersey	VOM, Diodes
RBM Controls 131 Godfrey St. Logansport, Indiana	Relays
Shurite Meters 428 Chapel St. New Haven, Connecticut	Meters
Superior Electric Co. 383 Middle St. Bristol, Connecticut	Connectors
Warner Electric Brakes Beloit, Wisconsin	Brakes

Appendix F
Mail-Order Houses

ELECTRONICS SUPPLY

Allied Radio Corp.	100 North Western Ave. Chicago, Illinois 60680
Lafayette Radio	111 Jerico Turnpike Syosset L. I., New York 11791
Newark Electronics Corp.	500 North Pulaski Rd. Chicago, Illinois 60624
Olson Electronics	260 South Farge St. Akron, Ohio 44308

SURPLUS

Airbourne Sales Co.	8501 Stellar Dr. Culver City, California 90230
Palley Supply Co.	2263 E. Vernon Ave. Los Angeles, California
Surplus Center	1000 West O Street Lincoln, Nebraska 68501

AUTOMOTIVE SUPPLY

Fredson Trailer Supply	814 North Harbor Santa Ana, California 92703
J. C. Whitney Company	1917 – 19 Archer Street Chicago, Illinois 60616

Bibliography

Chrysler Outboard Corp., Chrysler Corp., Hartford, Wisconsin.

Chrysler Outboard Service Manual, Glenn Marine Series; Cowles Books.

Clymer's Automotive Books (Various)

Crouse, William H., *Automotive Electrical Equipment*, 7th Ed., McGraw Hill Book Co.

D.A.T.A. Diode Book, Orange, New Jersey.

Ford Motors Service Manuals, Ford Motor Corp., Dearborn, Michigan.

General Motors Service Manuals, General Motors Corp., Lansing, Michigan.

IEEE Publication No. 45, "Electrical Installations on Shipboard", Institute of Electrical and Electronic Engineers.

Jeep Service Manuals, Kaiser Jeep Corp., Toledo, Ohio.

Motors Automotive Books (Various)

Outboard Motor Service Manual, Intertec Publishing Co., St. Louis, Missouri.

The Semiconductor Data Book, Motorola, Inc., Phoenix, Arizona.

Small Engines Service Manual, Intertec Publishing Co., St. Louis, Missouri.

Society of Automotive Engineers—1970 Handbook, Journal and Transactions, New York City.

Vinal, George, *Storage Batteries*, John Wiley and Sons, New York City.

Index

A
Alternating current	43
Alternator	113
Amperes	41
Ampere hours or seconds	42
Auxiliary power units	123

B
Batteries,	
primary	57
secondary	57
Battery	42
Battery capacity	65
Battery charging	62
Battery charging wires	86
Battery condition, measuring	58
Battery, dual	40
Battery loads	69
Battery maintenance & repair	63
Battery, second	31
Battery water	64

C
Capacitor	41
Cells	57
Class 1A	15
Class 1B	16
Class 2	16
Class 2A	16
Class 2B	17
Class 3	18
Class 4	18
Current	42

D
Diagrams, electrical wiring	52
Diode isolators	24
Direct current	43
Direct paralleling	19
Dropout voltage	54
Dual battery	40

E
Electrical loads	93
Electrical measurements	50
Electrical terms	41
Electrical wiring diagrams	52
Electricity, measuring	46
Electrodes	42
Electrolyte	42
Electromagnetic force	41

F
Farands	41
Field, magnetic	42
Fuses & circuit breakers	89

H
Henrys	42
Hertz	43
High voltage output	79
Hydrometer	58

I
Induces	42
Inductor	42, 43
Isolating relay of contactor	20
Isolators, diode	24
Inverting	43

K
Known & unknown factors	48

L
Law, ohms	47

M
Magnetic field	42
Magnet, permanent	43
Manual switch	19
Measurements, electrical	50
Measuring battery condition	58
Measuring electricity	46
Mechanical rectifier	45

O
Ohms	41
Ohms law	47
Ohmmeters	51
Ohms per volt	50

P
Permanent magnet	43
Potential	41
Potential or voltage drop	41
Primary batteries	57
Principles of operation	75

R
Rectifier, mechanical	45
Resistance	41
Running lights	101

S
Secondary batteries	57
Second battery	31
Series	42
Specific gravity	58
Switch, manual	19
Systems checks	70
Systems, twin engine	30, 34

T
Terminals	42
Terms, electrical	41
Trailer brakes, electrical	105
Twin engine systems	30, 34

V
Voltage	42
Voltage, dropout	54
Volt, ohms per	50

W
Watt-hours	42
Wattmeter	46
Wires, battery charging	86
Wire sizes	81
Wire terminals	102